Contract
Engineering

Contract Engineering

Start and Build a New Career

Amanda G. Watlington

Roger L. Radeloff

McGraw-Hill

New York   San Francisco   Washington, D.C.   Auckland   Bogotá
Caracas   Lisbon   London   Madrid   Mexico City   Milan
Montreal   New Delhi   San Juan   Singapore
Sydney   Tokyo   Toronto

Library of Congress Cataloging-in-Publication Data

Watlington, Amanda G.
   Contract Engineering: start and build a new career / Amanda G.
Watlington, Roger L. Radeloff.
     p.   cm.
  Includes index.
  ISBN 0-07-068498-7
   1. Engineers—Vocational guidance. 2. Contractors operations.
3. Engineering contracts. 4. Self-employed. I. Radeloff, Roger L.
TA157.W38 1997
620'.0023—dc21
                                     97-13036
                                         CIP

McGraw-Hill

A Division of The **McGraw·Hill** *Companies*

1 2 3 4 5 6 7 8 9 0   AGM/AGM 9 0 2 1 0 9 8 7

ISBN 0-07-068498-7

*The sponsoring editor for this book was Harold B. Crawford, the editing super-
visor was Penny Linskey, and the production supervisor was Sherri
Souffrance. It was set in Palatino by Victoria Khavkina of McGraw-Hill's
Professional Book Group composition unit.*

Printed and bound by Quebecor/Martinsburg.

McGraw-Hill books are available at special quantity discounts to use as
premiums and sales promotions, or for use in corporate training programs. For more information, please write to the Director of Special Sales,
McGraw-Hill, 11 West 19th Street, New York, NY 10011. Or contact your
local bookstore.

 This book is printed on recycled, acid-free paper containing a
minimum of 50% recycled, de-inked fiber.

Contents

Preface

1. Contingent Hiring in Today's World of Work   **1**

American Business Undergoes Fundamental Change   4
The Manager Assumes a New Role   9
Engineering: The Ideal Virtual Organization   11
A Call for New Strategies   13
Summary of Key Points in This Chapter   17

2. Writing a Résumé and Getting Hired by a Contingent Staffing Firm   **19**

Gathering the Needed Information   26
Focusing on Skills and Accomplishments   28
Guidelines for Successful Résumé Writing   36
Selecting the Best Résumé Format   43
The Rest of Your Marketing Package: The Cover Letter   50
Successful Interviews: An Active Approach   53
Summary of Key Points in This Chapter   59

3. What to Expect If You Work for a Staffing Firm   **61**

Applying for a Position   65
Making a Good First Impression   65

Halfway There   67
Leaping through the Next Hoop: The Contracting
Firm   70
It's All about Money and Such   74
Summary of Key Points in This Chapter   77

4. Making Contingent Employment a Career   **79**

Taking a Proactive Approach   82
Nontraditional Employment Tactics   93
What about Being an Independent Contractor
or Consultant?   95
Summary of Key Points in This Chapter   98

5. When Are Contingent Employees the Answer?   **99**

When Are Contract Staff the Answer?   102
It's Not without Drawbacks   103
Outsourcing and Other Options   106
Outsourcing Pitfalls: Beware of Losing Sight   108
The Leased Employee: More Control   109
The Independent Contractor: A Legal Morass   112
Summary of Key Points in This Chapter   115

**6. Teams, TQM, Work Groups: Integrating
    the Contingent Hire**   **117**

A Changed Context   118
Managing within a Difficult Context   121
Enhancing the Positive, Driving Out the Negative   123
Having Your Policies and Procedures in Place before
You Hire   127
Keeping Your Contract Employees on Track and
Performing   131
The Legal Obligations of Too Much Inclusion   134
Teams, TQM, Work Groups: Yet Another Set of
Challenges   134
Summary of Key Points in This Chapter   141

7. Selecting a Staffing Provider   **143**

The Value of Planning   146
The Account Executive: A Combination of
Professionalism and Chemistry   148

Reputable Firms As Providers of Reliable Staff   150
Staffing Firms As Service Businesses   152
Making Your Choice   155
Summary of Key Points in This Chapter   156

8. Developing a Strategic Alliance with a Contract Staffing Provider   **157**

The Value of Interdependence   159
Seeking an Alliance for Contract Engineering Staffing   163
Negotiating the Contract: The Provisions and Considerations   167
Getting the Support You Need from the Firm   171
Summary of Key Points in This Chapter   173

9. How to Get the Contract Employee You Need   **175**

Every Candidate Is Equal to the Sum of the Parts: Training and Experience   176
Looking Beyond the Job Description or Summary   179
Building the Job Order: Objectives, Tasks, Skills, and Experience   182
Refining the List: Tasks, Skills, and Experience   184
Clarifying Cultural Factors for the Job Order   185
Submitting the Job Order   188
Summary of Key Points in This Chapter   191

10. Evaluating the Contingent Hire   **193**

More Than Just an Annual Review   195
Phase I: Setting the Expectations   196
Phase II: Evaluating the Progress   199
Phase III: The Final Evaluation   203
Summary of Key Points in This Chapter   204
Resources   205
Resource Guide   209

Appendix   229
Index   241

Preface

From the author information, it is easy to assume that this volume is the collaboration of an ivory-towered academic and the president of a contract staffing provider, individuals both personally insulated from the reality of the changing employment paradigm. Although I have been a college professor teaching business and marketing for Terra Community College in Fremont, Ohio, since 1993, my own career has been shaped by the changing employment paradigm.

I am by nature not a risk taker, but I have been forced by unstable times to adapt and focus on staying employable. As a marketing professional in high tech, I have personally witnessed the dislocation created by the changes that have overtaken the world of employment. One former employer that enjoyed more than a 25 percent share of its market in 1978 no longer exists. As technology and the market changed, the firm underwent changes traumatic for its employees, including myself. Another company sold the subsidiary that I worked for, and my job evaporated. I have also had the dubious honor of having been downsized (along with 130 other people in my division) by voice mail. There was no in-person notice of the end of my job. There were just final instructions given on the telephone by a manager whose own job was also eliminated.

I pondered: "How could *all* these unpleasant experiences happen to me?" As I have discovered in developing this book, my experiences are far from unique. This book is not the musings of an academic; it is a survival guide written in part by a survivor. Its goal is to give guidance for the individual trying to stay employed in an environment that places the responsibility for career management clearly on the individual. It also is designed to provide that manager, forced to manage in a sometimes hostile environment, guidance in how to both profitably and humanely manage contract employees.

Amanda G. Watlington

Acknowledgments

The list of those to whom I am indebted extends from those who encouraged me to pursue a career in business to those who more recently provided support during the writing of this book. There is, however, one individual who has been there from the beginning—my husband, Mal Watlington. He has always offered encouragement and unflagging enthusiasm. No matter what direction my career has taken, I have always been able to count on his support.

In developing this book, I have had the pleasure of collaborating with a special individual, my coauthor, Roger Radeloff. Without his experience with engineering and the staffing industry, this book would not have been a reality.

I have also enjoyed the support of numerous individuals during the writing of this book. Special thanks are owed to my colleagues at Terra Community College, most notably Tim Gocke and Steve Uzelac, who provided encouragement at times when the demands of teaching seemed to overwhelm the process of book writing.

Thanks are also owed to Teresa M. Arnold and her associates, Brandy Baxter and Kari Bucher. Without their support and encouragement, this book would not have been possible. Then, there are the other professionals who provided information and guidance. The staff, clients, and other individuals associated with ITS Technologies were always ready to provide their insights and experiences. They unfailingly answered my questions. Also, special thanks to Karen Feld of the human resources group of Cinergy, in Cincinnati, Ohio, who provided information for this book.

It is my hope that every reader of this book will take away some kernels of information that will help each of them as he or she moves into the exciting world of the future. Today, we are all freed from the bonds of an unspoken but implied employment contract that we may find was in fact the proverbial pair of golden handcuffs. Although it provided some apparent career stability, it may have stifled career growth. These are exciting times that we each must seize.

Amanda G. Watlington

Preface

This book is written to provide insight to a changing world of employment and what you can do and how you can prepare yourself and enhance your employability for the future. The employment paradigms of the past are gone forever: Gone is employment for life. Gone is the idea of working for one company for life, being loyal, working hard, and earning a gold watch at retirement. The paradigm is no longer employment; instead, it is now employability. Today, and in the future, you need to know how to manage your own career and your employability.

Technical organizations need to better manage their resources and *unfix* their fixed costs. Successful, profitable technical services companies need to be very fast on their feet, respond to customer needs, and reduce costs quickly. When the customer's requirements change or when they demand specific services, you need to meet these demands. Contract services can be a key to meeting the demands for quick response along with the need for a flexible workforce.

This book will provide new insight to possible ways of managing your people, that is, your resources and increasing your overall efficient use of those resources. This book also considers how you can be more efficient and effective in meeting your customers' needs and how you can increase your competitive edge.

Roger L. Radeloff

Acknowledgments

I would like to acknowledge Dr. T. L. Cliff, Dr. Robert Avery, and Mr. Al Sanborn, all successful in building organization structures that provided excellent services along with providing opportunities for their employees to develop their careers and expertise in the professions. These gentlemen have had very different styles and approaches, but with similar characteristics: They have all had vision, have worked hard and have had the ability to direct others and provide the tools for success. They have also expected and demanded quality and productivity from all and from the organization. These high standards have resulted in high-quality products and services and have set them apart from their competition.

My wife Deanna, a successful and talented academic with an entrepreneurial spirit, has always given much advice and encouragement no matter how wild the idea. She never questioned, but always encouraged with support that allowed me the many opportunities that I pursued.

The person who made my publications and this book possible was Amanda Watlington, coauther. Without her ability, my thoughts and ideas would not have been put into words. She brings many ideas, experience, and insight to our changing world. What the future will bring to our world is exciting and stimulating.

Today, I am blessed to have associates whose values and goals are similar to mine. They range from recent graduates to those with experience greater and more wide ranging than mine. This creates an environment that is very exciting, challenging, and rewarding. I wake up every day looking forward to new challenges and opportunities.

Thanks again are owed to many people in my career who nurtured, encouraged, and provided the needed expertise. I hope this book helps to break down the old paradigms and provide new insight to employability and the ability to manage one's own career and to better manage organizations to compete successfully in a changing world environment.

Roger L. Radeloff

1

Contingent Hiring in Today's World of Work

*John, a midcareer civil engineer in Texas, was let go by a
large oil company just as oil industry employment started
to soften. Wanting to bring up his family in Texas, he
sought and found a new local employer outside the oil
industry. He thought he was fortunate and would avoid
further employment woes induced by the oil industry.
Little did he guess that less than a year later, he would be
looking again for a job. His new employer had experienced
a downturn in work related to the oil slowdown and its
effects on the overall Texas economy.*

*Both times John was laid off his employers told him how
much they regretted letting him go. John's performance on
the job was very good, and both layoffs were driven by
economic conditions. John has since held two more jobs.
With his children getting ready to go to college, he
frequently worries about his career and what has gone
wrong.*

The world of work has changed. During the recession at the start of the 1990s, the media carried daily announcements of plants closing and major corporations shedding thousands of employees. General Motors alone has downsized its workforce by 99,400, 29 percent of its workforce, and the two giants, IBM and AT&T, have eliminated 245,000 jobs, over 30 percent of their respective workforces. These pink-slipped workers, like John, were not all assembly-line workers. In the past, corporations responded to cyclical economic changes by adjusting their "blue-collar" assembly workforces. The jobs lost in the late 1980s and early 1990s were different. The pink slips were often issued to highly skilled technicians and those in corporate areas previously immune to layoffs.

In many instances, the workers were notified that they were not part of just a temporary staffing adjustment but a permanent reduction in force. The thinning of the middle-management ranks and the large number of layoffs and reductions in force among technical personnel were a signal that a profound change was rippling through the American workplace.

While middle management received the media hype about its dissolving and thinning ranks, the same major change was occurring in all areas of engineering. Engineers too were being let go in large, but unheralded, numbers. Many highly skilled engineers and technicians found themselves having to rethink what had once looked like stable careers. John's career instability has a happy ending. He has always found a new job. The rebounding economy and the still slow growth in permanent jobs have shown that the change is permanent. The permanent jobs just aren't coming back.

American business has rethought its manpower deployment strategies. Many individuals are unable to respond because the new corporate strategies do not mesh with their understanding of how they are expected to relate to the world of work. The change has caught them off guard.

Today, engineers and other skilled professionals must aggressively manage their own careers. Career experts advise that American workers will change jobs between five and seven times during their working lives. Engineers must take initiative and create opportunities for themselves or be prepared to ride the economic tide wherever it takes them. Every engineer needs to seek projects and employers that will further his or her career. It is no longer prudent or even sensible to take a passive role in career management.

There are a number of strategies for career self-management. A record number of individuals with management experience have chosen self-employment as a career strategy. Although the statistics on small-business failures are daunting, many former corporate middle managers have chosen to trust their own business skills instead of the vagaries of corporate life. Women in particular have chosen the self-employment option. Their situation is more compelling because of the still-in-place glass ceiling that seems to prevent women from rising to the top of the corporate ladder. Although there are signs that this phenomenon is changing, legions of women have chosen to trust themselves as entrepreneurs rather than wait for their turn in corporate life or beat against the glass ceiling. Not only does small-business ownership provide ample opportunity for personal growth for both men and women, it also presents equally large risks.

Others, more often engineers or those with highly specialized skills and deep connections within community or industry, have chosen to sell their services as consultants. Their former employers are often the source of their largest and sometimes only contract. This is an illusion of freedom. The consultant is like the proverbial man sleeping next to the elephant. If the elephant rolls over, the man is crushed. Unless the individual is prepared and has the skills to expand and market his or her services, the single-client contract is but a thin thread holding the former employee on a tether.

The individual, once fully connected to the corporate information pipeline, quickly becomes just enough outside the loops to be caught off guard by inevitable corporate shifts. There is also the added problem that corporate changes—moves, retirements, and reassignments—can shift those most instrumental for maintaining the contracting relationship. The contracting individual loses the glue that held the relationship together. As it loosens, so too does the ability to secure more work.

Some engineers and other technical employees have found that working on a contingent basis for a contract employment firm offers job security and increased opportunities for skill enhancement. These engineers have evaluated their options and chosen to use the contract staffing firm as a marketing agent for their skills and talent. In the future more engineers and technical personnel are expected to choose this employment option.

A few short years ago, engineers chose contingent employment as a way station between jobs. They could use their skills and earn a living while seeking their next permanent employer. There has always been a corps of engineers, particularly those in the defense industry, who have chosen to make contingent work their career. This book will examine why contingent employment today is often a career option of choice, not just a strategy for filling a career gap.

Since contingent employment is a different employment paradigm, it requires different modes for securing employment. Given the fundamental changes that are occurring in the national employment picture, managers can expect to manage growing legions of contingent employees. This will require adjustments for the manager. This book will examine some of the challenges this presents and offer guidance for this task.

This book is also a guidebook for engineers coping with the changing workplace. Unlike other books that have chronicled the phenomenon of the current workplace and simply pointed out what is happening, we will provide real information on how to seize control of your engineering career in a changing workplace.

American Business Undergoes Fundamental Change

The reality of the dramatic change in the workplace really set in when the recession ended and the economy started to grow. The jobs did not reappear. Instead of adding new permanent employees or recalling laid-off workers, organizations chose to use part-time or contingent workers. Today, contingent workers represent just less than 20 percent of the workforce. Experts predict that if the current trend continues, within 10 years, a third of all individuals will be employed on an as-needed basis. In 1993, temporary employment went up 21.3 percent over 1992. The temporary staffing industry is one of the fastest-growing industries in the United States. This is a sign!

This change in the shape of the workforce mirrors a fundamental change in how organizations of the future will view their structure and commitments. The British management scholar Charles Handy has suggested that the organization of the future could well be represented by the three-leafed shamrock. The first leaf would consist of the professional managerial core of the organization without which it could not continue to function. These core individuals would bring to the organization skills that are essential for the business's well-being. For example, a law practice could not function without attorneys, but it could very easily function without a bookkeeper and accountant, for their skills could be obtained from another source without jeopardizing the firm's ability to perform—that is, to provide legal counsel to its clients.

As the model presented above suggests, the second lobe of the shamrock would consist of those who provide the services to the firm as external contractors. The attorneys might choose to hire an external accounting and bookkeeping firm or other such support. Previously, every organization felt it needed to maintain its own support service corps. As competition has increased and the ability to rapidly connect with specialized services has improved, there is a lessening need to provide these functions internally. Further, as business has evolved, the skills needed to perform and manage these once so-called support systems have dramatically increased, thus making it more difficult for the internal managers to handle. Into this modern business environment, contract workers will often bring with them a very specialized and valuable set of skills. The organization will value the skill set and will expect to secure a positive cost-effectiveness from the relationship.

By contracting out for the services, managerial distraction is reduced. The manager can focus on the core business and can bring greater value to the organization. The contractor will be expected to ease the core business's managerial stress by providing a seamless fabric of services that will support the core business's ability to generate revenue.

The third lobe of the organization of the future will consist of contingent workers. These workers will include many highly skilled individuals, such as engineers and managers, who will be hired on a short-term basis to augment the existing business, whether it is a supplier (contractor) to the core business or the core itself. These contingent workers provide very specific, short-term support for the organization in times of expanded need.

Although represented as separate lobes of a single leaf, the workplace of the future will see fewer distinctions than this format suggests. For example, an individual may start a career within a core business, segue to the temporary lobe for an interval and be hired away to the contracting lobe, only to move again to one of the other lobes as his or her career progresses. This change in the fundamental structure of how we think of work is already being borne out in many organizations.

Many employers have already found that the model works for them. They have discovered that they enjoy maximum productivity when they operate with a core group of regular employees whose skills are critical to the business and supplement that group with additional workers on an as-needed basis. They have found that they can indeed readily add and subtract employees to meet changing demands without creating excess stress on the organization.

In 1992 William H. Davidow and Michael S. Malone published *The Virtual Corporation*, in which they described this phenomenon and the changing model for the world of work. They labeled "virtual organizations" those operating with only a small core of essential employees attached to the core business. In them, all employees besides the permanent staff are brought in as needed to produce whatever the organization needs. In virtual organizations, it is assumed that the employment of workers is contingent on the work that must be done. This means that when the task for which the individual is needed is completed or no longer necessary, the individual will be severed from the organization.

A growing number of companies are adopting this strategy or hybrid variants, and the three-lobed structure of the workplace is coming into sharper focus. Many companies and engineering firms are unaware that they are becoming virtual organizations. It is often a gradual shift—a single department is outsourced. The success of this results in other structural changes, and the organization becomes more and more virtual and more in tune with the new employment model. These organizations are simply responding to a changed marketplace. Competing in a world economy has forced manufacturers to reconsider all economic inputs—labor, capital, raw materials, and entrepreneurial talent.

Labor costs in the United States for many years exceeded other developing nations so strategies that reduced labor costs were initially very attractive. In competing with some countries such as Japan and Germany, the difference in wage rate is no longer the underlying key to competitive advantage. A company's ability to maintain a competitive advantage is tied to its productivity and more efficient use of the other manufacturing inputs, which in turn affect the company's ability to cost efficiently produce products and to meet or exceed the customer's expectations—the contemporary keys to advantage.

To bring a product to market rapidly requires a flexible approach to product development. Firms now seek speed and have achieved it through the use of temporary project teams. These teams are developed and staffed to meet the needs of a specific project. Individuals are chosen for the project because they have the specialized mix of talents and skills needed for the project's success.

Computer technology has simplified the management of interconnected project teams and the workforce overall. The development of technology is a key driver for the elimination of the middle manager and is the cause of some of the changes in engineering hiring. As a group, corporate middle managers are exceedingly vulnerable because of their role in the organization of the past. Their original role was direct supervision. They were also a liaison between the hierarchical layers of the organization. Their job was to smooth the pathway for information and decisions to transit throughout the organization. They were planners and deployers of individuals. They were essential.

Today, voice and data telecommunications networks can link entire far-flung units into a single seamless organization. Decisions can be instantly disseminated through voice or electronic mail systems, and an individual's work is readily available on the computer for immediate review even in another city. As a result, the need for the middle manager as a conduit is gone. Technology has made the manager redundant. Further, with increased emphasis on teamwork and improved customer relations, there is a driving need to flatten the organization. The middle manager is obviously the layer to remove since the middle manager no longer has a truly tenable role.

The middle manager in engineering and business often had a broad skill mix. This made it possible in the past to fit individuals interchangeably into a variety of management positions without the organization's developing yawning capability gaps. In the rapidly changing business environment, the middle manager has been forced to change roles to survive.

As their ranks thinned during the recession, the remaining middle managers had to expand their areas of responsibility. Many also found themselves called upon to perform high-level staff activities as well as manage. Today, a middle manager is often a content expert in a single area with a broad band of responsibilities in allied areas. In engineering, where technical expertise is the individual's franchise, the ability to combine strong content skills with management expertise is essential. This change has created a need for the manager to function like a peg— with skills that can anchor an area yet hold a work group together with overall management expertise.

These managers are part of the core of the business. Their content skills are essential for the organization. They must assume new roles that span the boundaries of the organization. They must be willing to look outside of their own walls in staffing their projects and understand far more about the labor market than they ever had to in the past when most of the project resources were internal—that is, captive resources, or known quantities that also carried a considerable knowledge of the organization's infrastructure and corporate culture.

The forces affecting management have already had profound effects on how all employees relate to their employers. The contract between employee and employer has fundamentally changed. In the past an engineering graduate could expect to secure a job and begin working on a recognizable career path. Some engineers would choose to make their career path within a single organization, accumulating years of seniority that at one time were viewed as worthy and valuable by the organization. Others would follow a more varied path directed by ambition, economics, or other factors.

One of the most crushing aspects of the recent wave of corporate downsizings for many individuals is how readily corporations have shed employees with many years of service. This is because, although the individual was unaware of a dramatic shift in the employee-employer work contract, a shift had indeed already occurred. Their severance notice was their first notification of a changed workplace. Others, like John at the beginning of this chapter, are still grappling with understanding the change while living and adapting to it.

In the past, business valued loyalty and having trained workers available. Today, every employee is an expensive and expendable entity. Corporate America is now constantly looking at each employee's productivity and ability to continuously add value. Many firms have justified moving to a nonsmoking environment based on data about work time lost to smokers and smoking. Employee wellness programs are sold on their records for keeping employees healthy and working. Every lost work hour has a value. Every hour of suboptimized productivity can transfer into lost profits. Underutilizing or overpaying employees for what they do is insidious and is no longer tolerable a form of suboptimization.

Businesses today hold their employees up to a different value measure than in the past. The employee's value to the firm is measured by the employee's ability to add value to the corporation at all times. To add value, the employee must have current skills and fit the organization's immediate needs. When an employee ceases to add value for the employer, there is no reason to continue the relationship. Even though an individual is no longer able to add value for one employer, the same individual may be extremely valuable for another. Instead of building a career with a single employer, the employee is forced to continuously sell his or her skills in an ever more demanding free market.

This need to continuously add value to the corporation has changed the premise on which engineers should base their careers. No longer are employees even remotely entitled to a lifetime, or so-called stable, job; instead, each individual must look for ways to maintain lifetime employability. The engineer who stays employable can expect to stay employed. This statement may seem tautological and overly simplistic, but it is the answer. For the engineer there is more to staying employable than simply jiffying up the résumé with each layoff. This book will address what an engineer must do to stay employable and offer strategies for achieving it.

Not only has the contract between employees and their employers changed, so too has the meaning of losing one's job. In the rip-roaring growth years from the 1950s to the early 1980s, individuals who changed jobs frequently or lost their jobs were often held as suspect. Even when the release was recession driven, prospective employers always had to consider why the individual had not been retained. They felt that there just had to be a reason. Individuals were cautioned to avoid a "spotty" résumé that might flag them as a potential poor hire or problem employee.

Employers were also wary of the "job-jumper"—those individuals who readily changed jobs and developed lengthy work histories. They were either viewed as problem employees or extremely ambitious and skilled individuals who could easily be lured to some new environ. Their loyalty was suspect. Career guides in the 1970s and 1980s were filled with cautions on having a résumé that showed no real commitment or progressive work history. Some aggressive job-jumpers heeding this caution would review their progress and seek ways to continue jumping, but less visibly. They would choose to change titles and work areas in a large corporation rather than move to another employer and risk the danger of a spotty résumé. A prospective employer operating under the old paradigm would scan an engineer's lengthy résumé for signs of a potential employment problem.

Today, employers must review in a different light a résumé listing a number of employers. Many individuals have experienced career changes over which they have had no control. For example, their entire department or division may have been severed in a single reduction in force (RIF), or they may have been caught up in a merger or acquisition resulting in hundreds of layoffs. In some vulnerable industries, an individual may have secured new employment just to reexperience the same career nightmare with a new employer. Prospective employers must now filter out the job seeker who fits this engineer's model from the true misfit.

The Manager Assumes a New Role

The ranks of the middle manager have thinned, and the role of the manager is evolving. Traditional management theory holds that the manager plays three roles: informational, decisional, and interpersonal. With computer and telephone technology speeding information throughout the organization, the manager must focus away from the informational role to the others. This does not imply that the informational role is diminished but rather that it will require less of the manager's daily effort to fulfill.

With fewer middle managers sharing the burden, today's manager must make decisions across a broader spectrum of activities than in the past. The staff resource planning and decision-making function itself has grown in importance. Today, the manager is no longer just confronted with how to deploy staff in a stable department but is also forced to continuously consider the validity of every staffing decision.

Using contingent labor in some way eases the hardships associated with making these difficult staffing decisions. Trying an employee on a contingent basis before hiring allows the manager to determine fit in real time. A person can also be brought in on a contingent basis to fill an evolving position. This relieves the manager from having to predict the direction in which the position will develop. If it fails to develop, the manager is not left with a disappointed employee or the problem of potential redeployment or termination. If growth takes the position in an unanticipated direction, the contingent employee can be replaced with an individual who brings the skills that the evolving position then requires.

A manager's interpersonal relationships with staff are also changing. In the past, successful managers sought to build a cohesive staff blending personal and organizational loyalty. Individuals seeking growth in the organization would form beneficial alliances with individuals in management who might further their careers. This was a mutually beneficial arrangement. Where there are large numbers of contingent or even part-time staffs, the equation changes. The criteria underlying individual loyalty are different. Today, the loyalties are shorter term and may be based on a contingent employee's seeking to ingratiate him or herself for contract continuance or on a part-time employee's looking for full-time work.

When the employee knows that he or she has been engaged on a project basis, the terms of employment are clearly stated. This often lessens the manager's grip on the individual and results in a more adult relationship that is based on open and clearly defined rules.

The management literature is rife with tracts decrying the need for empowered or self-directed individuals. In a world of work that stresses teamwork, the employee must focus away from meeting the specific, sometimes ego-driven, demands of the individual supervisor. The individual must be able to identify activities that will promote the good of the cause and help the team and ultimately the organization meet its mission. In our current employment paradigm, the individual is expected to develop new self-directed guidance systems for use on the job. These same guidance systems should be applied to the individual's career development process. The individual can no longer expect to have a benign employer forwarding the engineer's career. Even when the individual has a place on a manpower planning chart—the organizational team's depth chart—there is no guarantee that its promise will be fulfilled.

Followers of professional sports teams are seldom surprised when they hear that a team has waived, traded, or released a former superstar with diminished skills. The need to try out for the team and maintain a specific roster level is part of the idiom of sports. So too are the free agents. Many of these elements from sports are creeping into the workplace in general. Companies readily let go those employees with dated or diminished skills. Team chemistry is increasingly important in the business milieu, and the inability to work in a team will often guarantee release. Many companies are using strategies that let them carefully evaluate employees before they hire them permanently. Some select new employees from contingent or part-time hires just to ensure fit.

Free agency has driven up wages in sports. Will it do the same in business? We contend that to an extent the engineer can anticipate a marketplace that will value the professional's skills and will expect to pay for them in increasing amounts. A lot of this presumption is driven by market-related factors. In the future, the demand for intellectual skills and technological expertise will grow. Today, the value of the high school diploma has dramatically decreased because industry is no longer just hiring a set of hands to perform a task but is rather hiring a combination of higher-order skills. Engineers' stock-in-trade is higher-order technological skills. The price they bring in the market will reflect their scarcity and currency.

As more employees work on a contingent basis, performance evaluation will take on a new overtone. Since the individual is expected to add value to the process, the individual must develop skills that will meet specialized needs of the organization. Performance measurement will focus on criteria different from simply getting the job done. You can expect it to take a more global perspective.

Engineering: The Ideal Virtual Organization

Engineering is ideal for the application of the virtual organizational model. Most engineering is project or problem based. Throughout their schooling, engineers are taught to seek answers to specific problems, develop clear processes, or build clearly defined projects. Many engineering projects have obvious beginnings and ends.

The virtual organization and shamrock models easily adapt to situations in which the time frames are clearly articulated and the task requirements outlined—engineering. Many organizations, however, are a full jump or so from being ready to assume that *all* their employees will stay with the organization only for a brief while. These are the all-too-familiar firms that rapidly expand and contract their so-called permanent workforce to meet the demands of new projects. In the future, this costly process will diminish as more organizations realize the costs of this hiring and releasing process can be avoided through the use of contract or contingent labor.

In engineering, most tasks are skill-based. Although an engineer may receive a broad education in a specific major—for example, chemical, civil, electrical—it is the individual's work experience that will in the end define the person's skills and problem-solving capabilities.

Today's workplace relies on complex and highly specialized technology. The ever-increasing level of technological support for engineering functions is driving increasing levels of specialization into engineering hiring. For example, before the advent of CAD systems, technicians had to understand the complexity of the design requirements, but they did not also need proficiency in a specific and often highly specialized version of a software program. Although the software has increased the efficiency and productivity of the trained users, it has also increased the level of specialization required.

Now, an employer must recruit an individual with the design know-how *and* a specific level of proficiency with a specific software package. This branching into increasing specialization because of technology will continue. This is also a prime driver for contingent hiring. A highly trained individual with very specific software and problem-solving skills can be brought in to meet a specific engineering need and then be released when no longer necessary. Given the difficulty in finding individuals who have the required mixture of skill and technological experience, organizations may not even seek to recruit them. To secure these highly specialized individuals, the organizations will instead rely solely on agencies.

Engineering is also ideally suited for contingent or virtual hiring because engineers as a group prefer to work on defined projects or specific problems. Engineers when polled about their job preferences will often state that they want work that represents an interesting challenge. Many will choose challenging work above other traditional job satisfaction measures—advancement or even increased compensation. The virtual organization model presents the engineer the opportunity to move from one challenging project to another. For many this employment situation is ideal. As the concept of a changed work environment stabilizes and individuals and organizations consciously recognize that all work is project-based and that there is no such thing as a permanent job, it will be easier for individuals to reconcile their own feelings about what a job should mean and their need for interesting projects.

Given the nature of engineering training and hiring, compensation for engineers is often based on skill competency. This means that individuals can expect to be rewarded for what they can do, not just for their longevity with the organization. Skill-based compensation has been a reality in engineering for a number of years and readily transfers to the virtual model of the 1990s. This ability to receive appropriate compensation for highly specialized skills often makes it attractive to the individual to work on a project basis instead of molding away in a job that does not allow him or her to grow or use these valuable skills. Making this into a career, however, requires rethinking career planning.

A Call for New Strategies

This new work environment will call for individuals and employers to develop new strategies. The engineer will need to reconsider what a job really is. In the 1990s many individuals are being forced to understand and live with the reality of an impermanent job structure. Initially, this is a painful pill to swallow. It causes the individual (just as John did at the beginning of the chapter) to ask, what went wrong? The answer to "How did my career get so offtrack?" is not necessarily with the individual but is rather in the changed employment structure.

The sooner the individual comes to terms with the changed structure of work, the easier it becomes to develop a new personal strategy for obtaining more work. That is what this book is about. Understanding the new world of work and then how to relate to it are essential for success. With a new strategy in hand for obtaining work, the individual can then develop a revised career path. The hardest part is mourning the death of employment as it has been known in the past.

The engineer is not the only one caught in this spider web of change. For the manager, there is also a new set of rules. The manager will have to get better at determining staffing requirements. For every task, the manager must be able to understand and articulate the specific skills it will require. This will be increasingly important, because in the future the manager's role will be to ensure that the appropriate skill matrices are brought to their projects. The organization will be too lean to cut slack for miscalculations. A mistaken understanding of the skill requirements for a project will render the project unsuccessful or create delays that in turn will reflect on the manager.

The manager must also cope with the added dimension of finding individuals who can work within the culture of the organization. The increased use of teams and the need for individuals to work interconnectedly with others on projects has highlighted issues of culture in organizations. Fit is ever more important now than in the past where the manager could serve as a filter if the fit was imperfect. Now, where individuals must work well with one another without the manager, this filter system is gone. Careful hiring is the new filter.

The manager will also be charged with developing a virtual staffing model that will allow the organization to continue meeting rigorous standards imposed by ISO 900X or other such initiatives. Today, anyone coming into a total quality–driven organization without an understanding of the language and methods this entails will be severely hindered in performance. The manager will have to ensure that these initiatives stay ontrack while continuously integrating new, temporary employees into existing project teams. This is a tall order.

Many managers dread annual performance reviews as much as the employees they are forced to review. Some put this task off until they receive outside urging from their human resources departments. In the changed world of work, the evaluation of the employees' work will be ever more important. In those situations in which a person has been employed on a contingent or project basis, the decisions are very cut and dry. The evaluation of performance will not carry with it any record or organizational memory of previous good deeds done that can be used as a bank against which an individual can draw to offset the impact of temporary poor performance. The focus shifts to "What did you do for the organization today." It is a more ruthless evaluation than in the past.

In some organizations, it becomes important not to let a good contingent employee with a valuable set of skills slip away to another project or contractor. The manager wanting to retain the employee will need to plan ahead. The manager must know that a need exists or when it will arise, plan to keep the individual, and communicate this decision. This will require management to hone the systems that support human resource planning decisions. They will no longer need to focus entirely on the long-term issues of succession planning but will need to bring their scope in closer to project-level human resource planning. The line manager—not just the human resource manager—will need to develop more sophisticated skills in determining short-term needs.

The same increased technical specialization that is driving many of the fundamental changes in the world of work is making it ever harder to recruit employees. A technical recruiter must understand both technology and recruiting. As technology becomes more complex, the depth of the recruiter's knowledge has had to increase. Human resources has not escaped the corporate fat whacker's ax. These areas too have been greatly reduced as organizations have lowered their overall headcount. Many organizations have focused for several years on reducing headcounts, not on recruiting. This depleted area is now being asked to continually restaff the organization with a new type of employee.

To meet the challenge, many firms have turned to contractors, the specialists in staffing. Since they are constantly placing employees with contractors, they have highly sharpened recruiting skills. Their core competency is recruiting, evaluating skills, and matching individuals to job orders. These contract staffing firms also fit with the virtual organization model. They are experts at supplying labor.

With the increased use of contingent staff and the expanded role of the staffing firms that provide them, engineers should consider how these organizations might fit with their career plan. Every individual, loathsome though the task may be, is forced to self-market every time he or she seeks a new job. By developing a skill package and a self-marketing strategy that uses the contract staffing agency as a career agent, the engineer can develop a new career path. With an understanding that a career is a series of jobs with no expectation of real longevity, the individual can begin to determine what will transfer into long-term employability.

In the past individuals could rely on the corporation to fit their career development into the scheme of their department or work area. Today, the individual must take more responsibility for lifetime learning and developmental planning. In a new order where hiring is directly related to experience and skills, choosing ways of obtaining valuable skills and experiences will be the mode for future advancement. No longer will individuals advance because they hitched their career wagon to a rising corporate star. Individuals must be the star in their own right.

Engineers and technicians in software-dependent areas must seek to become not just conversant but fluent in each new version of the software. The software industry is currently moving at approximately a new version every 2 or so years. The rate is even faster in emerging technologies. This means that all individuals that use software are on a continuous learning curve. This book, for example, is being written on a software version that is less than a year old and represents the authors' sixth version.

A broader skill base will make the engineer more attractive to potential contractors and easier for the contract agency to place. The engineer with a vast array of skills and experience can become a star. Since many organizations using contract staffing are still struggling with how to articulate jobs in the new skill-based jargon, the individual with a broad skill base will adapt easier and find that contractors are more satisfied with their work.

A dossier of skills and experiences are required for success. Today, many new graduates complain that the only jobs open are those that require 2 to 3 years' experience. They ask, "How can I get experience if all the jobs require experience?" This difficulty has occurred because the world of work is shifting to skill-based employment. To respond to this, you must be prepared to outline and market skills, not just experience. Contract employment can provide the rapid advancement of skill deployment in a variety of industries. This will help the recent graduate develop the dossier of experience needed to secure the desired career opportunity.

The hiring manager must be able to determine an individual's skills and work adaptability immediately. Managers will read the individual's résumé with a different focus than in the past. They will spend less time looking at where the individual has worked and far more time looking at the individual's skills, what was accomplished, and the precise nature of the individual's role in the projects.

Contracting firms will become validators of the individual's experience. The manager hiring a contract employee will not have the time to continuously verify the skills and experience of every individual needed by a team or for a project. The contract staffing firm will be trusted to provide qualified personnel to meet the contractor's needs. Since these firms will play a significant role in the hiring patterns of the future, it is an important career strategy to know how to deal with these agents. Just as there are defined levels of trust in the employer-employee relationship, these same levels of trust will have to be developed between the engineer and the staffing firm. This is still evolving.

The easiest-to-understand model is that of free-agent athletes. Each of these highly paid individuals uses an agent to negotiate his or her contracts and to procure employment. The engineer in a contract employment market is very similar to the free-agent athlete. Just as an athlete will review a number of agents before choosing one to serve as his or her career mentor, so too must you. First, however, you must develop a document to help your prospective career agent evaluate your skills and to assist the agent in finding you contract work. You probably already have a résumé, but if you developed it for an employment paradigm of permanent work, you will need to rethink what you include.

Summary of Key Points
in This Chapter

1. American business has rethought manpower strategies, permanently altering the landscape of technical hiring.

2. The changed workplace calls for career self-management.

3. Contract employment through a staffing firm offers job security and opportunities, not just a way station between jobs.

4. The world of work in the future will consist of three main sectors: the core employees essential for the functioning of the business, a group of external contractors and vendors providing services that were once considered part of the organization, and legions of contingent or temporary employees brought in on an as-needed basis.

5. The development of "virtual organizations" is spurring the growth of "virtual employment."

6. The use of flexible staffing increases the ability of American businesses to compete in a world market.

7. Technology is fueling fundamental changes in the manager's ability to supervise and manage large numbers of employees.

8. To survive in engineering management, a manager must combine managerial expertise with content skills.

9. Employers value employees based on their ability to add value to the company.

10. Staying employable is the key to staying employed.

11. Managerial decision making today cuts across a broader spectrum of activities than it did in the past.

12. Engineers should transfer empowerment skills from the workplace to their career.

13. Increasing technological specialization is creating greater needs for technical experts and is driving the growth of contract employment.

14. New career strategies including self-marketing and maintaining a broad and current base of skills are key to keeping a career moving forward.

2

Writing a Résumé and Getting Hired by a Contingent Staffing Firm

Matt, a chemical engineer with several years' experience, works for a division of a company that has yet again not met its profit projections. He is sure that it is just a matter of time before his division is closed. The rumors are already flying that it will either be sold or closed within the year. Matt is concerned. He has very little seniority and does not anticipate receiving a large severance package. He also has very limited savings and knows that he must find another job quickly or face financial hardship. He is concerned that a flood of other job seekers from his own company will make it that much harder for him to get a job in his area. He decides that maybe he should seek a job through a contract staffing agency. He has heard that a number of firms in his area routinely use engineers hired through staffing firms. As a first step, he searches among his files and pulls out his résumé. He begins to work on it but asks: "What is the right format?" "How should it look?" "What strategies will help me get a job faster?"

When you send or electronically transmit a prospective employer a résumé, you are sending a snapshot of yourself and your career. A good résumé and cover letter can effectively market you and your career to employers. These documents are your most important marketing pieces. Even a good résumé will not net you a new position. It will open the door to an initial telephone contact and perhaps an interview. The telephone contact and interview are your personal sales opportunity. For a successful job hunt, you will want to sharpen your offering in these areas.

Whether you are applying to a staffing firm or directly for a permanent position, you should follow the same basic principles of résumé construction. Your résumé must be easy to read, attractive, and provide the information a prospective employer will need to evaluate you. All employers quickly skim through stacks of résumés or review a number of electronic submissions for any available position. Many spend less than a minute evaluating each résumé. They select for further consideration only those that jump off the pile or screen. You must seek to have your résumé stand out. To do this, you must present the details of your career in an easy-to-read and attractive format.

When you develop a résumé to send or submit to a staffing firm, you should also take into consideration a number of elements that are specific to this employment mode. A staffing firm typically will be hiring for a variety of positions on a continuous basis. The firm must maintain a database of capable candidates from which to select those who might fill the requirements of each new job order. Most recruiters and contract agencies have sophisticated electronic filing systems to help them match prospective candidates with job orders. A staffing firm will manually screen your résumé, then electronically scan it into the firm's computer system or move it electronically into their database. The staffing firm's recruiters then depend on their software to search the database of scanned résumés for very specific capabilities and characteristics.

If you understand the steps a staffing firm will take in handling your résumé, you will find it easier to develop a résumé that will land you a job. A résumé submitted on paper must be visually attractive to make it through the initial manual screen. You should assume that your résumé will become part of an electronic database. You can expect to have the firm scan your résumé into its base. Therefore, in selecting paper and a typeface for your document, you will need to remember that scanning equipment works best with plain paper and clear, crisp type. Although colored paper may make it stand out from a stack of printed résumés, it often does not scan well reducing its value. Select a font size of 10 to 12 points. Avoid the temptation to use slashes, dashes, asterisks, and other unusual characters. These can create confusion for text-sensitive software that incorporates them meaninglessly into the search criteria. A sensitive scanner will also pick up the pattern on a vellum, laid, or textured paper. A good rule of thumb is that if you think your résumé would not fax well, then it won't scan well. If you are unsure about how your résumé will scan, try it on a fax machine. Most have a copy function, so you won't even have to dial a number to conduct the test. If you have access to a scanner, just try scanning your résumé, and you can test its scannability yourself. Do not send copies of faxed résumés or copies with any marks from excess copier toner on them. The résumé you send to a staffing firm should be a clean, clear original.

Many staffing firms have sites on the Internet. The resource guide at the end of this book gives you Internet addresses for approximately fifty staffing firms and other employment sources. Many of these sites include either a résumé template to fill in online or will accept an electronically submitted résumé. The Internet is revolutionizing recruitment. With its broad application by business and the increased access for individuals seeking employment, you can expect to find it playing a significant role in your job hunting. Whether you are looking for a permanent position or for work in the expanding contracting sector, you will need to understand how to use the Internet to your benefit.

The résumé templates given by prospective employers present almost no formatting challenges; however, you must have the details of your career and accomplishments firmly in order before you begin filling out the document. You can waste a considerable amount of time canceling the submission because you do not have all the data called for immediately at hand. Also, since each is somewhat different, it helps to work off of a data sheet as opposed to a stylized résumé. When completing an online résumé, be sure to proofread it meticulously before you hit the submit button. You should avoid the temptation to send it before you have had a chance to carefully consider what you have typed into the blanks. The résumé you submit online is just as important to your job search as the neatly printed traditional résumé sent in the mail.

Since the recruiter's software uses key words to search for candidates, your résumé, whether submitted by mail or electronically, will need to be "hot" with words and information that recruiters want. Since most recruiters' job orders call for specific skills and capabilities, you will want to make sure that your résumé highlights yours. To help you identify some of these, Exhibit 2-1 gives a list of current skills and certifications. Scan this list and be sure to include those that fit your capabilities.

Although there are countless books on developing a résumé, many were written before the advent of the computer scanning technology

Instructions: Use this list to stimulate your thinking as you build your skill inventory. It is by no means exhaustive, and your personal experience will often enable you to draw from several discipline areas.

Aeronautical/Aerospace Engineer:

> *Computer Skills (Languages, Software, Hardware):* C/C++, CAD systems, DYNA 3D, f-DEAS, FEA, FORTRAN, IBM Mainframes, Mathematica, MATLAB, MSC/DYTRAN, MSC/PATRAN, MVS/XA, NASTRAN, NOVAK 2D, Novell Networks, Open GL, ProEngineer, ProMold, SGI, Sun UNIX, VAX, X Windows.

> *Other Skills/Experience:* Coding, design/design analysis, stress mechanics, systems analysis, troubleshooting, thermal/structural analysis.

Biomedical Engineer:

> *Computer Skills (Languages, Software):* CAD Systems, C/C++, FORTRAN, MATLAB, motion analysis software, Reflex.

> *Other Skills/Experience:* Bioinstrumentation, biomechanics, circuit tracing, control systems, fluid mechanics, hardware/software integration, mechanical design, system design/analysis, thermodynamics, test design, troubleshooting.

Chemical/Environmental Engineer:

> *Computer Skills (Languages, Software, Hardware):* CAD Systems, ChemCad, FORTRAN, HETEX, MathCad, MATLAB, OPSIM, PICLES, PRONTO, SCFRAC.

> *Other Skills/Experience:* Chemical production, chemical recovery, design, risk assessement, system modeling (air quality, chemical transport, chemical recovery, etc.), thermal/chemical analysis, troubleshooting, waste management.

Civil Engineer:

> *Computer Skills (Software):* CAD Systems.

> *Other Skills/Experience:* Construction, design analysis, design/drafting, project

Exhibit 2-1. List of current skills and certification.

and the Internet. We would like to offer you some advice on developing a résumé specifically directed at those employers, recruiters, and contract hiring agencies that use electronic information retrieval systems. You will find that many of the tips that we give you in this chapter on developing a résumé for a staffing firm might apply to any job search. They will yield a stronger résumé for engineering career management

management, structural analysis, systems design.

Electrical/Electronics Engineer:

Computer Skills (Languages, Software, Hardware): Assembly, CAD Systems, C/C++, FORTRAN, Novell Systems, UNIX, VAX/VMS, X Windows

Other Skills: Coding, design/drafting, installation, instrumentation, process control, project management, quality assurance, software design, system analysis/design, system evaluation, system integration, system scaling.

Industrial/Manufacturing Engineer:

Computer Skills (Software): CAD Systems, MRP Systems, ProManufacture.

Other Skills/Experience: Cost analysis/reduction, control systems, drafting/design, installation, job evaluation, material handling, method analysis, plant layout/evaluation, process writing, standard data analysis, system design/analysis, tooling, troubleshooting.

Mechanical Engineer/Designer:

Computer Skills (Software): CAD Systems, MRP Systems.

Other Skills/Experience: Automation (all aspects -- planning, installation, operations, etc.), control systems, cost reduction, drafting/design, layout design, machining, manufacturing, material handling and procurement, material properties, mechanical design/analysis, process systems, system design/analysis, tooling.

Metallurgical Engineer:

Computer Skills (Languages and Software): CAD Systems, C/C++, FORTRAN, graphics software.

Other Skills/Experience: Corrosion protection, equipment skills (e.g. lathe, mill, press, grinder), failure analysis, heat treatment, machine design, materials selection/design, metal joining, polymers, specification control, technology development and implementation (e.g. welding technology), thermodynamics.

Software Engineer:

Computer Languages and Operating Systems: Ada, BASIC, Banyan, C/C++, CMS,

Exhibit 2-1 (*Continued*).   List of current skills and certification.

COBOL, DOMAIN, GOAL, IBM assembler, Intel assembler, HTML, JAVA, Jovial, Motorola assembler, Novell Netware, Oracle, POSIX, Solaris, Sun operating systems, Sybase, VADS, Visual BASIC,VMS, UNIX.

Microprocessors: ATAC, Intel, and Motorola.

Standards/Tools: Accelerator, Adagen, Aprobe, CADRE Teamwork, Interleaf, Logiscope, MIL-STD, Rational Apex, TCP/IP Connections, TQM, X Windows, Yourdon OOD.

Hardware: Apollo, IBM Mainframe and PC, Rational, Sun, Tolerant, Univac, VAX.

Other Skills: Programming, software analysis, software design, software integration, client server, enterprise computing, LAN/WAN development, troubleshooting.

Applicable to All Disciplines

Computer Skills: Lotus Notes, e-mail, Internet, Office Suites (e.g. MicroSoft Office, etc.)

Other Skills/Experience: ISO 900X, TQM, team-work, communication training.

Exhibit 2-1 (*Continued*).   List of current skills and certification.

because the résumé you develop will be directed toward presenting your skills, technical expertise, and capabilities. The progression of the businesses you have worked for is almost coincidental to these elements, unless of course you have worked for highly prestigious companies. With many of these companies' shedding numbers of employees in the last few years, the value of your experience with some of these respected giants will be clouded by the huge number of other available individuals with similar employment histories.

Since a résumé alone does not get you hired, we are also including information on how to interview (both by telephone and in person) with a contract staffing firm and with prospective contractors. These interviews are crucial for selling yourself and your capabilities. As you begin drafting the résumé, you are actually laying the groundwork for successful interviews. When you seek employment through a staffing firm, you will have to sell yourself through at least two or often three levels of interviewing. You must first interview with the firm often by phone and then in person. This is followed in most instances by an interview with the client, the prospective contractor. There may be more steps, depending on the firm and the job. (Exhibit 2-2 outlines the steps in the hiring process at ITS Technologies.)

You must be absolutely ready for the interviews by the time you begin

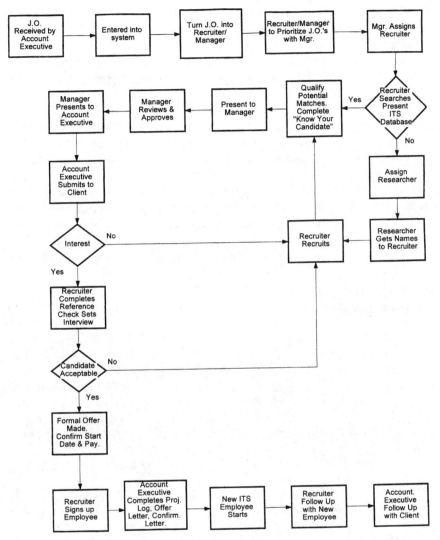

Exhibit 2-2. The recruiting work flow used at ITS Technologies, Toledo, Ohio.

to distribute your résumé. Employers depend on contract staffing firms to provide them with the needed employees on a timely basis. An employer places a job order with a contract firm when it needs someone. The optimal situation would include an almost instant connection with the exact right person. Any time spent searching and recruiting is time lost. This time has an inherent cost that both contractor and staffing firm would like to reduce.

Since the job order is placed with the firm when the individual is needed, there is little if any lag in making the hiring decision. When a contract firm sends them the right person, the hiring decision and work start date are almost immediate; therefore, you should also be ready to begin working almost immediately. The ability to fill positions on a timely basis is very important to the success of the staffing firm. The firm's revenues are based on placement numbers. The more rapidly they can fill the job order, the more quickly the placement can begin to provide revenues for the firm. An increase in staff working under contract to employers will increase the firm's revenues so there is no reason to delay. Given this scenario, you must do your homework and be ready to present yourself effectively to both the firm and its customers.

Gathering the Needed Information

Although you may already have a résumé, you should use the materials provided in this chapter to sharpen the information and its presentation. Most people hate to write their résumé. Take comfort in knowing that it is perhaps one of the most difficult career-related tasks that you must undertake. Don't flog yourself if you do not have an up-to-date résumé on hand. It is a rare individual who follows the sage advice of always keeping an updated résumé on hand. If you have spent a number of years with the same company and have not recently revised your résumé, you will definitely need to use the worksheets to gather the needed information.

The worksheet process of résumé development can be helpful whether you are an early career, recent graduate with limited experience to draw on or a seasoned veteran with a lengthy dossier. The data sheet development process will help an early career individual focus on elements that might enhance his or her limited work experience. The seasoned veteran is often confronted with an organization problem: how to distill a career into a few short pages. We have found in our experience that one of the psychological barriers to résumé development is the problem of distilling a long career. Sometimes the opposite situation exists: A person's long work history may actually be quite limited in its breadth and scope. Using the data sheet should reduce anxiety and make it easier to be objective in either case.

Using the data sheet will also help you avoid distorting your career record and help you increase your objectivity. It is easy to forget or distort elements of your work history if you begin writing without building a data sheet. As you work your way through a structured data gathering process, you may find that a job you enjoyed, because of the pleasant working atmosphere, was the least career enhancing. Because your experience with the job or employer was so positive, you might find yourself giving it more play on the résumé than it merits. On the other hand, you may have done your best work and grown the most on a difficult or even relatively short assignment. The structured information gathering process will help you filter your work history and put it in perspective.

Before you write your résumé, you should gather information on these five key areas:

- Work experience

- Skills and capabilities

- Education

- Military experience

- Professional certifications, civic activities, and personal accomplishments

Exhibit 2-3, the career worksheets, provides a guide to developing your résumé. You can make copies of these worksheets for your use. Complete each as fully as possible. The actual process of putting the material on the worksheet is very important. How many people have you met who have a great novel in their head, not on the best-seller list? The same goes for people with completed résumés.

Putting the data on the worksheets makes it real. As you work, follow the temptation to go back and add more detail. Let your mind hitchhike on information provided for one section to help recall data to fill in another. At this stage, don't worry about having too much information. This is a worksheet, not a final document. You will get a chance to edit and select the most important details later. Completing the worksheets will force you to focus more intensely on the important elements of your career. The threads around which you can build your résumé and your approach to your career will become more visible. Everyone's career has a story waiting to be told. Your résumé is your chance to tell your story. No one else will have exactly the same story. As you gather the information and build your résumé, you are telling your story.

A. Data on Your Work History

Complete one worksheet for each position you have held since completing your education. Copy as many sheets as you need. It is often easier to begin with your most current position and then work backwards. You will need less complete details about positions held earlier in your career if they led to more responsible positions with similar requirements.

Your Title _____

Years Held: From _____ to _____

Firm's Name and Address _____

_____

Industry and Description of the Business (Annual sales, size of engineering staff, etc.)

_____

_____

_____

Reported to (name and title)_____

Responsibilities (use short phrases with strong action words and give only general descriptions

that fit your position) 1._____

2._____

3._____

4._____

5._____

Special Skills or Software Required:

1._____

Exhibit 2-3. Career history worksheet.

Focusing on Skills and Accomplishments

All employers are looking for winners. Staffing firms are looking for employees who will thrill their clients and be versatile enough to work in a variety of situations. It is not in their best interest to obtain an

2._____

3._____

4._____

5._____

Accomplishments:

1._____

2._____

3._____

4._____

5._____

Reasons for leaving:

1._____

2._____

Exhibit 2-3 (*Continued*).   Career history worksheet.

employee who can fulfill only a single contract. The agency would like to keep and maintain excellent employees as much as the employee would like a steady job. When the agency recruits an individual, there are considerable costs associated with the hiring process. The agency looks to recover these costs through multiple successful placements of the individual.

If you possess either an impressive array of skills or highly specialized capabilities, do not hide them. If your career has been primarily in a major industry, you will want to be sure that your résumé reflects this depth. The staffing firm will find this information useful.

Throughout your résumé you will want to focus on your accomplishments. For greater impact and credibility, it is important to wrap numbers around your statements: "Developed library of 400 Autocad drawings." "Designed wastewater system with 2-million-gallon-per-day capacity." Don't worry about seeming to "toot your own horn." This is

B. Data on Your Education and Training:

Advanced Degree(s): Complete one sheet for each advanced degree.

    Degree (M.S., Ph.D.)_____

    Specialty Area _____

    School Awarding Degree and Address _____

    _____

    _____

    Year Awarded _____

    Fellowships/Scholarships Held _____

    _____

    Honors/Awards Received or Other Special Accomplishments _____

    _____

    _____

Advanced Degree(s): Complete one sheet for each advanced degree.

    Degree (M.S., Ph.D.)_____

    Specialty Area _____

    School Awarding Degree and Address _____

    _____

    _____

    Year Awarded _____

    Fellowships/Scholarships Held _____

    _____

Exhibit 2-3 (*Continued*).   Career history worksheet.

Honors/Awards Received or Other Special Accomplishments _____

_____

_____

College Degree(s):

Degree (B.S. or B.S. E.E.)_____

Specialty Area (Electrical or Chemical Engineering)_____

School Awarding Degree and Address _____

_____

_____

Year Awarded _____

Scholarships Held _____

_____

Honors/Awards Received or Other Special Accomplishments _____

_____

_____

High School

School and Address_____

_____

Year Graduated _____

Honors/Awards Received or Other Special Accomplishments _____

_____

Exhibit 2-3 (*Continued*).   Career history worksheet.

Professional Certifications (P.E. or C.N.E., etc.)

1._____

2._____

3._____

Specialized Training and Additional Courses Taken

    If you have taken any courses that might supplement or support your career, list them.

_____

_____

_____

_____

_____

Proficiency -- Computer Software and Hardware

_____

_____

_____

_____

Military Service

    Service Branch _____

Exhibit 2-3 (*Continued*).   Career history worksheet.

expected. There is a difference between providing a record of your accomplishments and "puffing" your career.

Puffery will catch up with you. Falsifying information on a résumé is grounds for termination virtually everywhere. If you go to work for a contract staffing firm and the contractor expects you to have certain

Year Entered _____ Year Discharged _____

Rank at Discharge _____

Type of Discharge _____

Special Training _____

Reserve Status _____

Accomplishments _____

_____

_____

C. Data on Your Professional/Personal and Civic Activities:

Professional Memberships:

National: _____

_____

_____

Local: _____

_____

_____

Publications/Papers /Speeches (Use additional paper, if necessary)

_____

_____

_____

_____

Exhibit 2-3 (*Continued*).   Career history worksheet.

Community and Civic Activities (membership, committees, and honors) _____

_____

_____

_____

_____

Exhibit 2-3 (*Continued*).   Career history worksheet.

skills that you have unrealistically represented, you can expect to be fired from the contract *and* by the firm. On a contract assignment, you will be expected to perform.

When you review what you did on each job, consider that some aspects of the job which you may have thought quite routine may actually be very valuable. As you complete the information, you will find that you are writing the story of your career. You should pay special attention to the story your résumé tells. Each position is a chapter, and you will need to explain why you closed each chapter. Job candidates who do not have a well thought out reason for leaving one job and going to another hurt their credibility during the interview phase. They seem unsure and often present a negative impression because it appears that they are hiding something.

If you harbor some resentment or still ache from the recent loss of your job, the worksheets can help refocus you on what you really did. Take comfort in your accomplishments. Writing them down will also help to bolster your ego. Health specialists consider losing a job one of the most stressful life passages. Many individuals find it difficult to write their résumé simply because it stirs up so many unpleasant emotions related to the loss of their previous employment. It is often hard to move on and ready yourself for the challenges of a new job—or the job of looking for work—if you have not worked through the emotions tied to the job loss.

Even if you are making the decision to move on without the strong prompting from your current employer of a layoff or other threat to your job stability, there is still a lot of emotional baggage that you will need to shed. You may find yourself questioning your judgment or the wisdom of your decision. Any vacillation or lack of clarity about the decision will translate into either a poor presentation on the résumé or into less than stellar interviews. Your heart just won't be into the process. With its inherent potential for rejection, you must do what you can to bolster your ego during the process. A clear career story with closure on each chapter will eliminate some of the sense of attack and add to your sense of personal well-being. You will not feel the sense of too many loose ends that can create tension.

It is critical to your success in obtaining employment that you are ready to move on. To do this, you must develop distance and the ability to interview without showing hostility or resentment toward your past employers. You should develop an account of what caused the end of your job that you can tell without hints of bitterness, anger, or even lingering sadness.

Once you have thought out how you will present the facts, you may want to practice telling your story until it feels comfortable. Try it out on your spouse or on good friends who know your situation. Ask them to tell you if it seems contrived or phony. Ask them to critique your body movements and tone of voice. Did you seem defensive? Did your voice seem artificially flat—too rehearsed? How did they react to your story? Being able to tell your story naturally will make it easier for you to interview and network.

As you work on your résumé and your career story, you should look for an emerging theme. For example, the theme may be that you are an excellent detail person, or a top-notch communicator, or the go-to person when big problems arise, or the dependable, never-complain, never-have-to-explain individual. When you find the theme, it will give you a thread on which to base your answers to the inevitable interview question: "Tell me about yourself."

By organizing your information on these worksheets, you will be more prepared to answer this and similar questions in the pressure-filled atmosphere of an interview. Since the interviewer will be working from your résumé, you can expect the questions to flow from this document. From your experience with the worksheets, you will be ready to answer questions without hesitation and provide requested details. You will not have to scour your memory as you grope to remember some significant point as the pressure mounts in the interview. The better organized you are, the more prepared you will be to answer most questions posed by an interviewer whose only previous contact with you is your résumé. You should always remember that you are the only person who can answer these questions:

- What were the challenges you faced in each position?
- What interesting dynamics did you face?
- What were the technical challenges?

- What positive results did you achieve?
- What are your strongest or weakest skills?

Once you have completed the career history worksheets, you will be ready to begin writing your résumé; however, we urge you to finish reading this chapter before you write your first draft. You will find the other information in the chapter helpful. Résumé writing is such a difficult and personal process that if you can achieve better results with less effort, you will be a step ahead. However, avoid the other temptation of procrastination. Just read the chapter and get on with the process.

Guidelines for Successful Résumé Writing

Your completed worksheets simply provide the data that you will need to write your résumé. You must now turn this career data into easy-to-read information. First, you must consider how you will present your career information. The following proven guidelines will help you avoid making common résumé errors.

Limit your résumé to two pages. Even if you have had a long and distinguished career, your résumé should still conform to this standard. Remember that your résumé provides a snapshot, not a photo album, of your career. Meeting this requirement will often force you to distill the important elements of your career into a few pithy sentences. Yes! It may be painful to have to explain 5 to 10 years of your work life in a single short paragraph. When your résumé gains you an interview, you can then expound in the detail you think your career merits.

Use action words to describe your accomplishments. For each position that you have held, you should develop an action statement of what you accomplished. You were a participant in the activities, so you should always frame your statements in strong active words. To help you accomplish this, you will find a list of action words for résumés given in Exhibit 2-4. If you start your descriptive sentences with one of these words, you will find that you are immediately in the action, not a bystander. Also, the county fair principle is well represented in résumé writing: "You must be present to win." If your résumé makes you look like a bystander or spectator in the activities, your role may be questioned.

Eliminate or minimize gaps in your work history. If you have had several short periods of unemployment, you may want to use years instead of complete dates for any chronology of your work history. This will help

Accepted

accrued

achieved

acquired

administered

advised

analyzed

anticipated

appointed

arranged

assessed

audited

augmented

authored

avoided

Began

bolstered

bought

Centralized

certified

clarified

classified

collaborated combined

composed

conceived

concentrated

concluded

conducted

connected

consummated

controlled

converted

corrected

created

cut

Decentralized

declared

decreased

defined

delivered

demonstrated

designed

determined

developed

devised directed

documented

drafted

Effected

eliminated

employed

enforced

engineered

enhanced

enlarged

established

evaluated

evolved

executed

Exhibit 2-4. A glossary of "hot" words to use in resumes.

minimize the gaps. They won't glare. Be careful to not falsify or extend the dates; just lessen the impact of the gaps. You should be prepared to coherently explain any lengthy gaps because any prospective employer will want you to account for them.

exercised

expanded

expedited

explained

extracted

Facilitated

fixed

forced

forecasted

formed

formulated

founded

framed

Gave

generated

guided

Headed

hired

Illustrated

implemented

improved

improvised

increased

influenced

initiated

inspired

inspected

instructed

insured

integrated

interpreted

interviewed

introduced

invented

investigated

isolated

Joined

Kept

Launched

lessened

liquidated

located

Made

maintained

managed

merged

measured

minimized

modernized

Negotiated

Observed

obtained

Exhibit 2-4 (*Continued*).   A glossary of "hot" words to use in resumes.

offered

operated

organized

originated

purchased

Passed

penned

performed

piloted

pioneered

planned

predicted

prepared

presented

prevented

probed

processed

procured

produced

protected

pruned

published

Readied

recommended

recovered

recruited

redesigned

reduced

referred

regulated

rejected

related

renegotiated

renovated

reorganized

reported

resolved

restored

reversed

reviewed

revised

revitalized

Safeguarded

saved

scheduled

secured

selected

separated

served

settled

simplified

simulated

sold

solved

sought

sparked

specified

staffed

standardized

stated

streamlined

studied

submitted

Exhibit 2-4 (*Continued*).   A glossary of "hot" words to use in resumes.

suggested

supervised

supplied

supported

surveyed

T abulated

taught

terminated

tested

traded

trained

trimmed

U nified

united

updated

upgraded

used

utilized

V aried

verified

W on

worked out

wrote

Exhibit 2-4 (*Continued*).   A glossary of "hot"
words to use in resumes.

Develop a career summary instead of career objectives. In a few positive
sentences, you should highlight your strengths and achievements. If
you state career objectives, you may close more doors than you open. A
prospective employer may read your objectives and determine, without
seeking further explanation, that his or her firm has nothing of interest

to offer you. Use lots of nouns and verbs in this summary. Some experts suggest that you should have at least 20 to 25 strong nouns in the summary. It should punch, not ramble along.

Include your name, address, and phone number(s) where you can be reached. This is a sticky area. If your current employer is unaware of your job search and you wish to protect your privacy, you may decide to not include a work phone number. If you do not own an answering machine, now is the time to get one. Be sure that you obtain one that you can readily check when you are away from home. If you are conducting a job search and use an answering machine, you should check it regularly. Similarly, if you do not have a personal e-mail address and a connection to the Internet, you should consider getting one. If you include a personal e-mail address, you should be sure to check it regularly.

You should be prepared to return a prospective employer's phone call or e-mail message promptly. This means you will also want to carry a calendar with you for telephone calls. Treat your job hunt as you would a job. If you are unemployed sending out résumés, booking and going to job interviews takes the place of deadlines, meetings, and other work activities. You should always be prepared to make an interview appointment. When a contract employer gets a job order, speed in finding an individual that will meet the client's needs is of the essence. You don't want to miss opportunities because you could not be contacted readily.

Do not include salary information. Your skills and experience will dictate your value in the marketplace, so salary information is usually not very meaningful. Unless you have some specific reason for including your salary history, you can deal with this either in your cover letter or in the interview. You should maintain, however, an accurate salary history so that you can honestly answer questions related to salary.

Do not include a list of references or letters of reference. You also don't want to write "references are available upon request." It is assumed that you will have references and can furnish them. Save your valuable résumé space for important information. Be sure that you have current information, names, titles, addresses, and telephone numbers for your active references. Be prepared to give this information out when it is asked for. You will want to put together a sheet of references that you can take to interviews and leave behind upon request.

Interviewers will sometimes find it helpful in checking your references if you include what your relationship with each reference has been. This is particularly important if your reference is a former colleague or boss who has moved to another company. You don't have to trace their genealogy, just give enough information to cue your reader about the individual's relationship to your career.

You should, as a courtesy, contact your references before you begin your search. When your reference is a former colleague or supervisor who has moved on to another position, you may find it useful to bring him or her up to date on what you have been doing since your last contact. This allows him or her to speak knowledgeably on your behalf. This is particularly important if you have enjoyed a promotion or achieved certification since your last contact. Your references should provide a balanced view of your work. If you were part of a team, include a team member or other peer. If you held a management or supervisory role, you should include both bosses and subordinates.

Avoid giving personal data. According to the law, you don't have to divulge your age, race, health, or marital status or the number of children you have. Although you may have fascinating hobbies, their mention is inappropriate. You are seeking a job with a contract firm, not a place on a game show. I remember one such inappropriate résumé wherein the writer indicated as a significant achievement for the year "catching a sailfish off the coast of Mexico." The job the individual was applying for had nothing even remotely related to fishing or any reason to include this tidbit. The ridiculousness of it has stayed in my mind, however, for years.

Proofread and check your résumé twice, and then go over it twice more. Go over your résumé very carefully. Don't just rely on the spell checker in a word processor to pick up all the possible mistakes. It won't catch a mistyped phone number or address. A single slip of the finger could change a date by 100 years or claim that you are proficient in a nonexistent version of software. Any errors will glare and could lose you a potential interview and job. This is not a time to rush. Be extra careful about phone numbers and addresses. Because we are most familiar with them, our eyes tend to skim over them.

Prepare your document and set it aside overnight. Then, in the clear light of a new day go back over it. Don't be afraid to revise it. Many people treat their résumé as if it were an epitaph that once written, is carved in stone forever. Your résumé is your personal marketing piece and merits refinement. After you refine it, let it sit again, even if just for a few hours. Then, reread it again as though it were written by and about a stranger. Sometimes when we revise a document, we will unwittingly disconnect elements that previously fitted together.

Make enough copies. This may seem trivial, but you will use more copies than you imagine. You will need copies to mail and to keep for use in interviews. Sometimes mailing and handling can batter even sturdy paper. It never hurts to be able to give a fresh copy to someone interviewing you. You want your résumé to present you at your best in all instances. Make sure that all your copies are high quality—no toner streaks or discolorations.

Mail it in a flat envelope. When you submit your résumé to a staffing firm, be sure to send it in a flat envelope. If you anticipate doing an extensive job search, you might consider buying a package or box of 9-by 12-inch envelopes. (Don't forget that these larger envelopes weigh more and will probably require more postage.) You do not want to send a folded résumé. First, it will not lie flat in front of a recruiter for ease of manual review. The recruiter will have to fight to keep it flat. Second, and far more important, the folds in the paper will create a problem for the firm's electronic scanning equipment. The fold marks may actually cause some of your vital information to scan poorly.

Truth-test your final résumé. Read your finished résumé over very carefully for accuracy. Do not claim education, experience or accomplishments that are not your own. When you are hired, your résumé goes on file. Falsification is grounds for dismissal.

Selecting the Best
Résumé Format

There are two typical formats for professional résumés: chronological and functional. In this book we are going to suggest that you consider a résumé style that blends elements of each of these formats. As you read along, you will see why this approach is suited to the special needs of the engineer trying to obtain employment in an alternative work situation.

The chronological résumé is the most traditional and still very popular. Exhibit 2-5, sample chronological résumé, is in a standard chronological format. Most job seekers use this format because they know that some organizations are still less accepting of the functional résumé. The standard chronological résumé, however, is more likely to bury an individual's skills and engineering-specific competencies.

Since we recommend that you use a résumé that offers the best features of the chronological and functional résumé styles, it is important to look at the specific features and characteristics of each. Using a chronological résumé style, you would include your name, address, contact information, and a career summary listing the positions you have held, beginning with your most recent and working backward. On it you will list significant accomplishments for each position. It is not recommended that chronological résumés include accomplishments from jobs more than 10 years in the past.

For older information, it is recommended that the candidate simply list

TERRY SMITH, P.E., CmfgE

---

243 West Avenue   Phone: 421-256-2232 (Home)
Apt. 2B   421-258-3254 (Work)
Toledo, OH 43431

Summary
Candidate has a range of experience in CAD, AUTOCAD ver. 12, and mechanical design. This includes plant and equipment layout, materials handling equipment design. Food processing and automotive industry experience. Additional skills as CAD instructor. M.S. in mechanical and B.S. in engineering physics.

Previous Work Experience

UTC Technologies, Inc.
Toledo, Ohio

CAD Designer   April 1995 - Present
Responsible for plant and equipment layout and electrical updates using AUTOCAD ver. 12.

ABC Design
Toledo, Ohio

CAD Designer   January 1994 - April 1995
Contracted to Engineering, Corp. Responsible for the design and detailing of food processing and material handling equipment using AUTOCAD ver. 12.

ABC Tube
Toledo, Ohio

Mechanical Designer   July 1992 - January 1994
Responsible for tube milling machine design and detailing. Work was done from Engineer's notes, sketches, and already existing drawings to complete projects.

Jones Community College
Toledo, Ohio

Instructor   August 1988 - Present
Responsible for teaching lectures and labs in AUTOCAD and manual drafting, manufacturing engineering technology courses concentrating in manufacturing management, plant layout, work design, process planning, and JIT. Advisor to several student organizations.

Exhibit 2-5. Sample chronological resume.

years of employment, the organization and titles without elaboration. Many individuals find it convenient to list these positions under "other experience." An interesting or meaningful part-time or graduate or summer internship that directly related to professional growth or other career enhancing opportunities are listed under "other experience."

Drafting, Co.
Toledo, Ohio

Senior Manufacturing Engineer   August 1977 - August 1988
Responsible for capital acquisition and maintenance, financial justification, capital forecasting, process improvements, process troubleshooting, and long range planning. Project leader to implement manufacturing changes because of several major design modifications to automotive gears and crankshafts.

Education

University of Toledo
Toledo, Ohio
M.S. in Mechanical Engineering/August 1976
B.S. in Engineering Physics (Mechanical Engineering option)/March 1977

Continuing Education Courses:

Quality Control	Gear Manufacturing
FMS Techniques	Induction Hardening
Heat Treating	Project Management

Software Knowledge:

AUTOCAD ver. 12	CADKEY
FactoryCAD	Quattro Pro
Wordperfect	Lotus

Exhibit 2-5 (*Continued*).   Sample chronological resume.

After the chronological treatment of the career, résumés in this format list the individual's education, beginning with the most advanced degree. Then, the standard chronological résumé ends with any significant military or professional or civic activities that the job hunter views as integral to his or her career.

The chronological résumé enjoys broad acceptance because it is so easy to follow and presents the individual's career in a logical manner. It is a suitable format for an engineer whose career has followed a logical progression of increasingly responsible positions. You should view it as a linear representation of your career. If your career has not followed a linear progression, the style adapts poorly. It is not well suited to those whose careers have been a string of projects requiring special skills but not following in a neatly linked progression.

The chronological résumé meets the needs of an old-paradigm worker who expected to have a career that moved along in a logical sequence of progressively more responsible positions. This style was also ideal for those who enjoyed a stable career with a limited number of employers, each representing a significant segment of the person's career growth. This style also gave individuals with long careers in prestigious companies an opportunity to showcase their accomplishments within this impressive context.

Although it does not enjoy the popularity of the chronological résumé, the functional, skill-focused résumé is more in tune with the changing world of work. Exhibit 2-6, sample functional résumé, shows how the same résumé might appear following a functional format. The average worker will make at least seven job changes throughout a career, and there is no data that suggest that the average engineer will be less mobile than other workers. Given the nature of engineering work, however, he or she may even exceed this expectation.

As work histories become more cluttered, employers have to evaluate candidates differently. They are focusing more on the individual's skill mix rather than on the companies the person has worked for. The functional résumé highlights the skill mix and the individual's accomplishments.

The functional résumé is very similar to the chronological résumé in that it gives the same initial directory type of information—name, address, contact information, and career summary at the outset. Then, the two types begin to diverge. The first page of a functional résumé is dedicated to outlining the candidate's skills and areas of expertise with accomplishments pinned under each area or skill. The second page is then given over to a summary list of the individual's work history. Functional résumés also end with the individual's military and professional or civic activities.

The functional résumé is usually used by individuals who want to make a major career change and need to highlight skills and accomplishments that will transfer from one career path to another. This type of résumé is also very useful for the individual who has a long history with one job and a lot of skills and accomplishments packed into one job. On a chronological résumé this would look like the writer had a severely limited work history; on a functional résumé, however, an individual's accomplishments would shine. A functional résumé also will minimize gaps that might create doubt and unwanted questions.

The functional résumé does not occupy the same level of acceptance that the chronological résumé enjoys. The poor acceptance of this type of résumé is due in part to the fact that it enables a candidate to minimize career liabilities. A contract staffing firm, particularly interested in

TERRY SMITH, P.E., CmfgE

---

243 West Avenue	Phone: 421-256-2232 (Home)
Apt. 2B	421-258-3254 (Work)
Toledo, OH 43431	

Summary
Nineteen years of mechanical design, CAD design , materials handling and engineering experience in food processing and automotive industries. Experience includes:

Plant layout and design	Capital acquisition and maintenance
Design Troubleshooting	Capital Forecasting
JIT	Robot installation

Accomplishments

- CAD design for new materials handing equipment using AUTOCAD ver.12.

- Converted engineers sketches and existing drawings for 10 tube milling machines.

- As project leader implemented major manufacturing changes due to modification to gears and crankshafts.

- Over 10 years of experience teaching CAD design using AUTOCAD and manufacturing engineering.

Previous Work Experience

UTC Technologies, Inc.
Toledo, Ohio

CAD Designer   April 1995 - Present
Responsible for plant and equipment layout and electrical updates using AUTOCAD ver. 12.

ABC Design
Toledo, Ohio

CAD Designer   January 1994 - April 1995
Contracted to Engineering, Corp. Responsible for the design and detailing of food processing and material handling equipment using AUTOCAD ver. 12.

Exhibit 2-6. Sample functional resume.

your skills and accomplishments, will appreciate the way a functional résumé highlights your skills. The chronological résumé, however, gives a clear and well-accepted overview of your career and is thus easy for a recruiter to follow. It is recommended, therefore, that you use a style that blends the best elements of each type of résumé.

ABC Tube
Toledo, Ohio
Mechanical Designer   July 1992 - January 1994
Responsible for tube milling machine design and detailing. Work was done from Engineer's
notes, sketches, and already existing drawings to complete projects.

Jones Community College
Toledo, Ohio

Instructor   August 1988 - Present
Responsible for teaching lectures and labs in AUTOCAD and manual drafting, manufacturing
engineering technology courses concentrating in manufacturing management, plant layout, work
design, process planning, and JIT. Advisor to several student organizations.

Drafting, Co.
Toledo, Ohio

Senior Manufacturing Engineer   August 1977 - August 1988
Responsible for capital acquisition and maintenance, financial justification, capital forecasting,
process improvements, process troubleshooting, and long range planning. Project leader to
implement manufacturing changes because of several major design modifications to automotive
gears and crankshafts.

Education

University of Toledo
Toledo, Ohio
M.S. in Mechanical Engineering/August 1976
B.S. in Engineering Physics (Mechanical Engineering option)/March 1977

Continuing Education Courses:
    Quality Control   Gear Manufacturing
    FMS Techniques   Induction Hardening
    Heat Treating   Project Management

Software Knowledge:
    AUTOCAD ver. 12   CADKEY
    FactoryCAD   Quattro Pro
    Wordperfect   Lotus

Exhibit 2-6 (*Continued*).   Sample functional resume.

The engineering skill résumé style recommended in this book will yield a résumé that is easy to follow, yet skill focused. Exhibit 2-7, a sample engineering skill résumé, shows a résumé prepared in this format. You will notice that in this résumé you prominently feature your skills. Short, telegraphic hot words and phrases highlight the skills and accomplishments. Then, the standard chronological format gives your work history. When you use this résumé style, balance your entries so that you can give enough information on valuable skills to trigger the contract firm's computer search software.

TERRY SMITH, P.E., CmfgE

---

243 West Avenue   Phone: 421-256-2232 (Home)
Apt. 2B   421-258-3254 (Work)
Toledo, OH 43431

Summary
Nineteen years of mechanical design, CAD design , materials handling and engineering experience in food processing and automotive industries. Experience and skills include:

AUTOCAD ver. 12	CADKEY
FactoryCAD	Anvil 4000
Plant layout and design	Capital acquisition and maintenance
Design Troubleshooting	Capital Forecasting
JIT	Robot installation
Quattro Pro	Design detailing
Wordperfect	Lotus

Previous Work Experience

UTC Technologies, Inc.
Toledo, Ohio

CAD Designer   April 1995 - Present
Responsible for plant and equipment layout and electrical updates using AUTOCAD ver. 12.

ABC Design
Toledo, Ohio

CAD Designer   January 1994 - April 1995
Contracted to Engineering, Corp. Responsible for the design and detailing of food processing and material handling equipment using AUTOCAD ver. 12.

Exhibit 2-7. Sample engineering skill résumé.

ABC Tube
Toledo, Ohio
<u>Mechanical Designer</u>   July 1992 - January 1994
Responsible for tube milling machine design and detailing. Work was done from Engineer's
notes, sketches, and already existing drawings to complete projects.

Jones Community College
Toledo, Ohio

<u>Instructor</u>   August 1988 - Present
Responsible for teaching lectures and labs in AUTOCAD and manual drafting, manufacturing
engineering technology courses concentrating in manufacturing management, plant layout, work
design, process planning, and JIT. Advisor to several student organizations.

Drafting, Co.
Toledo, Ohio
<u>Senior Manufacturing Engineer</u>   August 1977 - August 1988
Responsible for capital acquisition and maintenance, financial justification, capital forecasting,
process improvements, process troubleshooting, and long range planning. Project leader to
implement manufacturing changes because of several major design modifications to automotive
gears and crankshafts.

<u>**Education**</u>

University of Toledo
Toledo, Ohio
M.S. in Mechanical Engineering/August 1976
B.S. in Engineering Physics (Mechanical Engineering option)/March 1977

Continuing Education Courses:
  Quality Control   Gear Manufacturing
  FMS Techniques   Induction Hardening
  Heat Treating   Project Management

Exhibit 2-7 (*Continued*).   Sample engineering skill résumé.

The Rest of Your Marketing Package: The Cover Letter

Your basic personal marketing package will consist of your résumé and
a cover letter. Since you will want to apply with several contract staffing
agencies, you should develop a standard letter to send to multiple agencies. If you are working with a computerized word processing system,

it is recommended that you develop a template that you save. Resist the temptation to write individual letters. Craft a wonderful initial letter that will serve as a base for your job hunt. Exhibit 2-8 gives a sample base letter and shows how you might alter it to meet specific situations.

If you have a particular job title that you are applying for or know that the firm handles or has openings for a specific title, be sure to include the title in the letter. Many agencies advertise for specific job titles in both professional literature and local newspapers. This will ensure that the right person sees your documents.

To generate additional letters, you will want to be able to simply call up your base letter, alter it as you see fit and then save it under a new name. This will maintain the integrity of your original document so that you can use it over and over again. Over time you will develop a library of letters that use your cover letter as a base. Using a base cover letter makes it easier to remember what you included without having to resort to rereading the letter in the middle of the interview.

If you are conducting your search using e-mail messages, resist the temptation to dash off a quick note to accompany your résumé. Craft

Date

XXXXXXXXXXXX
XXXXXXXXXXXX
XXXXXXXXXXXX

Dear xxxxx:

As you search for professionals with experience in (*your specialty*), please consider my credentials:

(*Include three or four key elements from your resume -- your special educational attainment, your years of experience in specific industries, and your accomplishments in previous jobs*)

I would like to put my experience and expertise to work at a contract position with your firm.

I look forward to discussing my background with you in greater detail. Please feel free to contact me at (*your contact number*).

Sincerely,

XXXXXXXXXX

Enclosure (resume)

Exhibit 2-8. Sample cover letter for submitting resume to staffing firm.

your e-mail message as carefully as you do your cover letter. Use the same base letter with adaptation to fit the changed format. Construct your letter offline, and then submit the letter via your e-mail append function. This is very helpful if you are in an area that does not provide local phone access to your e-mail. You do not want to rush the letter writing process with a clock ticking on your minutes of access time or your phone bill. Just a word of caution on e-mail file submissions: Before you send any e-mail files or submit your résumé, you should check to see if there are any format restrictions that you need to be wary of. It would be tragic to miss the perfect job opportunity because your résumé was not in a file type read by the receiving agency's equipment.

It is suggested that you keep cover letters sent by mail to a single page. You are not interviewing by mail. You are trying to get an interview, wherein you can tell your story. With word processing software, it is not very difficult to create your own distinctive letterhead. It doesn't have to be fancy. A larger bold type of the same font will yield an attractive letterhead. Make sure that your name, address, and telephone number are on your letterhead. Infrequently, a résumé will get separated from the cover letter during the interviewing and hiring process. If your contact information appears on all documents, there is no way for it to get lost. Letterhead, even done on a computer, gives a visually more satisfying package than plain paper.

No matter how long you make your cover letter, be sure to sign each copy before you send it. This is a very common mistake made by users of standardized computerized letters. They get so busy cranking out the letters and putting them in envelopes that they forget to sign them. They even sometimes will forget to enclose the résumé or put a stamp on the envelope. Check all of these seemingly minor details. You surely do not want to send your résumé "postage due."

Under no circumstances send a standard letter that has been copied on a copier. No matter how broadly you are casting, you must bait each hook separately. Sending a copier letter is a tip off that you are sending a blanket mailing to all of the contract staffing firms. Not only is it obvious, but it looks like you are trying to take a shortcut on the process and makes you look lazy.

While you are at the computer developing a cover letter, you should also draft a standard thank you letter that you can send immediately after your interviews. If you interview quickly and heavily, you may find it quite difficult to get these very important notes sent out promptly. The thank you note is a courtesy that indicates not only that your manners are intact but also that your interest in the firm is high. If

Date

xxxxxxxxxxxxx
xxxxxxxxxxxxx
xxxxxxxxxxxxx

Dear xxxxxxxx:

I thoroughly enjoyed meeting with you to discuss the contract assignment of (*insert job title here*) with your firm. The assignment, as you described it, sounds challenging and of interest to me.

With my education and experience, I am confident that I could quickly become a productive part of your organization. I look forward to discussing further how I might contribute.

I look forward to hearing from (*account executive's name*) at (*staffing firm's name*) of your decision.

Sincerely,

xxxxxxxxxxx

Exhibit 2-9.   Sample thank you letter for staffing firm interview.

you do not send a note, the interviewer has nothing to reinforce any signals of keen interest that you might have given during the interview.

The more standard letters you set up before you start interviewing, the more prepared you will be to conduct an effective search. Since contract hiring moves quickly as the firm tries to fill a job order, you will need to be ready to respond with thank you notes. Exhibit 2-9 gives a thank you note suitable for contract hiring. One is used when you have interviewed with a firm. Another is used when you have had an interview with a prospective contractor. Use the letter only as a guide. Alter it to suit your personal style. A canned letter, borrowed from a book, is often quite obvious and will detract from your first impression.

Successful Interviews: An Active Approach

The key to successful interviewing is an active approach. When you get a call for an interview, you must be prepared to take charge of the situation without wresting it away from the interviewer. Many job seekers approach the interview from a passive vantage. They expect the interviewer to handle the entire situation—ask all the questions and guide the candidate along. The successful job hunter uses a positive interview approach to make a positive impression. Active interviewing shows a can-do, interested individual who will probably make an excellent employee. Contract agencies look for the self-directed individual who will be able to handle himself or herself assuredly. For more positive interviews, follow these guidelines.

Know who will be interviewing you. Each person you come in contact with will require a different strategy. Find out as much as you can about the person you will be interviewing with, and use the information in your preparation. It is recommended that you take notes to help keep the individuals sorted out. You may be asked to interview with several individuals between the contract firm and the prospective contractor. Don't just take down a name and address. Ask questions that will help you understand the individual and the position you are interviewing for.

Know the organization. Find out as much as you can about the organization. If you are interviewing with a staffing firm, you will want to develop a list of questions to help you evaluate whether the employment situation suits you. In Chapter 3 (pgs. 68-70), we will address this more thoroughly. When you are interviewing for alternative employment, you have to consider the opportunity from two angles. First, do you want to work for the firm? Second, do you want to work on the specific contract you are being considered for? If you do not seek information on both dimensions, you may find yourself liking the firm and not the contractor or vice versa.

Develop your personal presentation. If you have not interviewed for several years, you may be rusty. To work out the rust, you should practice answering job interview questions. Exhibit 2-10 gives a list of questions that you should review in your preparation. If you develop answers to these questions, you will not find yourself struggling for the right words in a stress-filled interview situation.

You should also develop how you want to present your skills to an interviewer. You will have to convince the staffing firm that you should work for them. Then, no sooner will you have done this interview than you must impress a contractor. You should be prepared so well that you can handle multiple interviews within a short span of time without losing your edge.

Anticipate the questions, and don't get caught off guard. If you have worked out smooth answers to the sample interview questions, you will not be caught having to ad-lib and risk answering a tricky question inappropriately. You will also be better prepared to anticipate the direction the interview is going and be able to respond. There is absolutely no substitute for meticulous preparation.

1. Tell me about yourself. (Develop a five-minute self-presentation to fit this type of question.)

2. What are your short- and long-range objectives? What have you done to accomplish them?

3. Apart from compensation, what do you expect to gain from this position and how does it fit your goals?

4. Why are you looking at contract employment?

5. What are your greatest strengths?

6. What are your greatest weaknesses?

7. Why should we hire you and make an investment in you?

8. What can you contribute to our organization?

9. What are your three most significant accomplishments and why were they successful?

10. What type of contract position do you think you are qualified for?

11. Why have you been successful in your career?

12. What is the most difficult decision you have ever made?

13. What is the biggest mistake you have made in your career?

14. What types of things do you enjoy and how have they influenced your development?

15. Describe your ideal position.

16. Describe your best and worst boss.

17. In what type of environment do you function best?

18. Tell me about a situation where you were ineffective?

19. Tell me about a few situations where your work was criticized.

Exhibit 2-10.   List of job interview questions.

Accentuate the positive. The active interviewer takes a positive spin on even negative questions such as "What are your weaknesses?" or "Describe your most difficult boss." These questions are deliberately asked to trap you into revealing something negative about yourself. We all have weaknesses, and we have all made career mistakes. You show your maturity and your ability to self-assess if you acknowledge your weaknesses and then describe either how you identified them or what

20. How would your co-workers describe you? Your subordinates?

21. How would you describe your own personality?

22. What didn't you like about your last position, boss, organization?

23. Give me an example of how you handled a difficult peer, boss, and subordinate.

24. Why did you leave your last position? (Answer this question briefly and without hesitation.)

25. What limitations do you think you have technically?

26. If people were to criticize you, what would they pick on?

27. What is the most difficult type of environment for you to work in?

28. What is the greatest risk you have ever taken?

29. What are you not as proficient in, or do not like to do?

30. What is the biggest disappointment you have suffered in the last 2 years?

31. Describe your communication skills.

32. What specific actions did you take in your last position to improve your effectiveness?

33. How do you stay technically current?

34. How did you spend your time at your last position?

35. How do you handle pressure?

36. What qualities do you look for in other people?

37. Have you ever worked in a self-directed work group or on a team?

38. Other than challenging assignments, what factors must be present in a position to keep you satisfied.

39. What is the most important thing you have learned about yourself in your career?

40. How do you set and meet priorities?

41. How do you and your family feel about relocation?

Exhibit 2-10 (*Continued*).   List of job interview questions.

42. What did you like best about your previous positions?

43. How do you define success?

44. What is the best advice/criticism you have ever received?

45. Have you ever considered owning your own business? What kind? Why?

46. Why are you interested in this agency?

47. What significant industry changes do you see occurring and how will they affect you?

48. How are you preparing for the future?

49. What are the most significant challenges you would expect to meet, if you were chosen for

this position?

50. What interests you the most about the position, and the least?

Exhibit 2-10 (*Continued*).   List of job interview questions.

corrective actions you have undertaken. Avoid beating up on yourself and seeming to apologize too much. You do not want to seem insecure or too prone to self-deprecation.

Another successful strategy is to describe your weakness(es) in such a way that the employer might see it (them) as strength(s). For example, if you are a perfectionist or an ever-ready volunteer, admit it and then note how you are tempering your trait. If you tend to be impatient, you can admit a "bias for action" that is sometimes misread as impatience. This acknowledges how other people might perceive you but also provides a strong positive basis for the so-called fault.

Prepare for the logistics. Know where you are going and how to get there. If you are driving and are unsure of the area, spend time with a map to plot your trip. Take traffic into consideration, and be sure to add extra time to allow you to come through any security areas. Bring extra copies of your résumé. You will want to have the résumé in front of you while you are talking to the prospective employer. When you are interviewing with a firm, bring your list of references.

Anticipate testing. The staffing firm may ask you to submit for drug testing or other psychological testing. Today, you can expect drug testing in most businesses. Some firms use psychological tests to help them determine whether you will fit a specific personality profile that their clients need. Don't let this intimidate you. They are simply trying to get a better fit. Try to act naturally, and answer the questions honestly, not defensively.

Panel interviews. You may be asked to interview with a group of people. These interviews are often used by organizations that expect you to work with a number of different people each representing a specific area or type of expertise. They are also used to develop rapidly an objective view of a candidate. Each individual gets to see your response to every other individual and can use this information in his or her evaluation of you.

Prepare for these interviews by making sure that you get the name and title of each person interviewing you. During the interview stay alert and sensitive to the group dynamics and try to identify the group leaders. As you answer each person's questions, be sure to make eye contact with others in the group. This will keep all of the group involved in the interview and in your responses.

Critique your performance. After each interview, review what you think went well and what needs improvement. Try to assess what you learned from the interview so that you can continue to develop your interview style. As a contract employee, you will probably find yourself interviewing often. Before you go to a new contract, the contracting company will want to get to know you through an interview. The more natural and poised you become during the interview process, the easier it will be to sell yourself to prospective employers or contractors.

Do your follow-up immediately. Follow each interview with a short thank you note. Include any agreed-upon next steps. If a contract firm sent you on the interview for a prospective contract, you should get in touch with your staffing firm contact immediately after you leave the interview. If at all possible, you will want to talk with the account executive before the potential contractor makes contact. Your account executive is interested in placing you and can often help you. Information is the currency of this transaction.

When you contact your account executive, you can expect to be asked for your assessment of how the interview went. Be honest, and let the account executive know what you think were the high points and any potential pitfalls. Did their expectations as outlined in the interview mesh with those given you by the account executive? The communication between you and your account executive is very important. The account executive is your career lifeline. You will want to be sure that you advise the account executive of any next steps that you discussed with the prospective contractor. The employer will be contacting the account executive almost immediately after interviewing you and will be discussing these next steps with your account executive. If the interviewer left you with any hints as to the expected direction, you might find it useful to verify your impressions of the process with your account executive.

With your marketing package completed, your interviewing techniques readied, and your job hunting strategies defined, you are prepared to begin the actual process of looking for a job with a contract staffing firm. Although each firm will follow a slightly different pattern, you can expect a reasonably similar process. You will need to understand the process and how to evaluate a job offer for contract employment.

Summary of Key Points in This Chapter

1. A résumé and cover letter are the job hunter's key marketing tools. They must be outstanding.

2. Contract staffing firms maintain electronic databases and search for candidates using sophisticated software that matches key words with their résumé database.

3. Contract staffing firms use electronic scanning equipment; thus, to be highly effective, all materials should conform to this technology's special needs.

4. Electronic submission of résumés is changing how recruiting is done. Prepare your own résumé package to take advantage of the Internet.

5. There are two basic styles of résumés: chronological and functional. The engineering skill résumé, which we recommend for contract employment, is a hybrid that blends the styles to highlight the engineer's specific skill mix.

6. Interview actively. Engineers working in the contract environment can expect to interview frequently. They should be prepared for the interview both mentally and physically.

3

What to Expect if You Work for a Staffing Firm

Matt prepared his résumé and began getting his personal marketing package in order. As he finalized his letters and practiced his interviewing techniques with friends and family, he found himself deciding, "Should I apply to a number of staffing firms all at once or just one at a time?" He knew other engineers who had applied to several staffing firms at once, but he wasn't sure if this was the right approach to getting hired. Matt's personal financial picture would not allow for a protracted job search, and he wanted to land a job quickly. He worried, "How quickly will I know if a staffing firm is interested in me?"

The most pressing issue for Matt was determining what sort of a work life he could expect. "Would contract employment offer benefits and vacations?" He and his family always enjoyed their vacations, and Matt certainly hoped that alternative employment would not render these niceties a memory. "Would he be able to save for his retirement?" Until his division's poor performance cast doubt on the future of his job, Matt had really enjoyed his work. His employer offered good health, retirement, and vacation benefits. He had really appreciated the work atmosphere and found himself worrying about how he might find career satisfaction in contract employment.

Getting hired by a contract employment firm mirrors the process familiar to those who have sought permanent positions—résumés are submitted, interviews conducted, references checked, tests taken, and
offers tendered. The only thing different about seeking this type of work
is that instead of working at a single company for a single employer, you
will be the employee of a staffing firm and will probably find yourself
working for a variety of different businesses while under the firm's
employ. The type of work, the length of the assignment, and the actual
work life itself will vary depending on the staffing firm and its clients.
It is therefore important that you carefully evaluate the staffing firm and
the opportunities it offers before going to work. In this chapter, we will
discuss how, when, and what you should consider so that you might
find each experience as a contract employee a career-enhancing opportunity instead of a way station between jobs.

To improve your skills at evaluating a contract opportunity, we are
going to detail the roles of the employee, the staffing firm, and the contracting business (that is, the staffing firm's client). As a contract
employee, you will be in a three-way working relationship. Traditional
employment is only two-way, wherein the employee and the employer
determine their mutual benefit and compatibility during the preemployment process. Contract employment interposes a third party—the
contract staffing firm. Working with this third party, as a strategic ally,
you can create a stronger, more stable working environment than you
could as an individual. This requires, however, making a more comprehensive preemployment analysis. You must carefully evaluate the contract staffing firm and the contract. For a successful working relationship to flourish, there must be mutual benefit for all three parties.

Your first step to success is to place your résumé in front of the
recruiters of a firm that understands your qualifications and can mesh
them with the needs of an appropriate client. Your chances of finding
the right contract staffing firm can be improved if you take advantage of
the screening conducted by a professional association focused on making good employee-employer matches happen. This book includes a
resource guide of staffing firms. Many are members of the National
Technical Services Association (NTSA), Alexandria, Virginia. The
staffing firms listed in the resource guide as members of this national
organization are committed to serving the needs of both their clients *and*
with providing technical contingent employees opportunities for professional growth within the industry.

The firms listed as NTSA members are members of a national trade
organization that specifically targets technical services, that is, engineering. These staffing firms cover a broad technical spectrum and offer a

wide range of technical services to industry and government. The members are leading providers of contract technical staffing. They specialize in providing highly qualified designers, drafters, engineers, computer programmers, systems analysts, and other key technical employees. These staffing firms routinely serve the aerospace, electronics, automotive, and other industries heavily dependent on technical talent.

The technical staffing industry is rapidly growing, and today, it generates $5 billion a year. NTSA estimates that its member firms employ more than 200,000 technical personnel. For the job seeker this represents a significant zone of opportunity. By obtaining employment through a member of a national industry organization, you are assured of working for an firm that is committed to technical staffing and to the industry. The ability of the firm to keep you working and the integrity of the firm are important considerations, as you consider contract employment as a career.

National Association of Temporary and Staffing Services (NATSS), another industry organization with a broader focus and also located in Alexandria, Virginia, estimates that in the last 10 years, the temporary workforce has increased from 400,000 people to more than 2 million workers, or 1.5 percent of the workforce. Comparing these numbers with those provided by NTSA, you will notice that technical staffing represents a subset of the greater temporary employment market. It is important to realize that, as an engineer with valuable skills, the contract world for technicians and engineers is distinctly and subtly different than that of the semiskilled temporary clerical or manufacturing employee.

As more contract employees fill jobs requiring increased technical and computer skills, employers are discovering the value of their contract employees. Many of the positive features enjoyed by engineers and technical contract staff are slowly moving across the broader landscape of contract and contingent employment. Contract technical employees have typically held assignments significantly longer than other such workers. They have never been the temporary hired for a few weeks to fill in while permanent employees are on vacation or when manufacturing demand has outstripped existing staffing levels.

The entire face of temporary staffing is changing. As more businesses move to using flexible staffing to create competitive advantage, they are keeping the temporary staff longer. Instead of viewing them as temporary fill-ins, their temporary staff and other contingent workers are part of their manpower planning. The statistics show that in general, 56 percent of all temporary assignments last more than 11 weeks with 11 percent lasting more than a year. These statistics apply across the entire scope of temporary staffing, not just technical staffing.

No longer are companies willing to spend the time to familiarize a so-called temporary with the organization's workings just to have to repeat the process in short order. Today, companies are more willing to maintain long-term staffing flexibility by keeping their contract employees on longer contracts. These employees are valued members of the organization.

You can anticipate that the majority of the potential contracts you might be offered will not be "short-short" term positions. Companies are today not only using contract personnel for longer assignments but they are also providing them a work life that more nearly parallels their own employees. As you send your résumés, be prepared to embark on a new type of career. It will most probably consist of a number of positions of varying lengths. Some may be a few weeks or months, but as the industry matures, there are more documented accounts of contract workers with multiyear experience with the same firm on the same contract.

When you select a contract staffing firm, you should consider the possibility that it will more than likely become your long-term employer. As the demand for contract employees continues to grow, this will place increased pressure on the contract staffing agencies to recruit and retain employees that their clients find acceptable. With this pressure from their clients, the staffing firm will worker harder to retain their skilled staff. An indication of the pressure on the recruiting side is already seen in the large number of available contract jobs advertised in a variety of media. A few hours on the Internet will give you a good picture of the opportunities available. We highly recommend that you visit some of the Internet sites given in the resource guide at the back of this volume to familiarize yourself with the changing world of contract employment.

Given the large number of available opportunities and the increasingly permanent nature of contract employment, you should carefully evaluate the staffing firm as well as any potential contract you might be offered. As you develop your career in contract employment, the quality of the staffing firm or firms that you choose to work for will impact not only your ability to obtain work but even the work itself. An agency that has a strong reputation for providing high-quality, highly skilled engineers will receive the contract that includes the high-skill requirements. These are often the best-paying jobs. A respected agency also has a broader marketing appeal through which to obtain additional contracts for its employees to fill. They get the job orders in part because of their success in filling job orders quickly and with satisfactory personnel.

Applying for a Position

With résumés (both paper and electronic) ready, letters prepared, and interview etiquette in order, it is now time for the applicant to look beyond the mechanics of preparing to apply for a job. We recommend that you submit your résumé to a number of contract staffing firms. Some companies may use multiple staffing firms to fill a single position. Your résumé may seem more attractive to one recruiter than to another for even the same position. Sometimes, several staffing firms will scramble to find just the right individual to fill a very specific position. The staffing firm that finds a suitable employee first basically wins the placement. This is one of the reasons you should be ready to go to work immediately upon receiving an offer.

In the previous chapter, we detailed some interview strategies to help you get ready. You can anticipate that your first interview will probably take place over the phone. If the recruiter assigned to fill a client's job order thinks you might be the right person based on the résumé you submitted, the first step will be the telephone interview. By phone, the recruiter will probe more deeply to verify his or her first impression of you. If the recruiter has any questions on the details of your résumé or experience, the telephone interview questions will address them. You can expect questions that are directed to your skills and experience. The recruiter probably has a job order sitting in front of him or her and is probing to see how you match the requirements stated by their client.

The recruiter will also want to go over details of your résumé and develop a clearer picture of the individual behind the résumé. Many successful long-term placements are the result of the individual's ability to fit the culture of the client's organization. A résumé rarely gives a clear picture of an individual's ability to fit. Based on your résumé and your answers to the telephone interview, the recruiter will determine if you should be interviewed more thoroughly for possible presentation to the firm's client.

Making a Good First Impression

If you receive a call from a recruiter at a firm to whom you have sent a résumé, keep in mind that this is often the first level of applicant screening. Your résumé has jumped off the stack on the recruiter's desk, and now it is your turn to sell yourself. Although these telephone interviews are mostly informational, don't forget the old adage "You never get a second chance to make a good first impression." Treat this phone call with the recruiter as your first and, therefore, key interview. It is not a casual phone call.

If you have submitted résumés to contract staffing firms and anticipate receiving telephone calls in response to them, you should prepare yourself for the phone interview. You should also consider how to handle the logistics of the interview. If you are unable to talk at length at the exact moment the call comes, you should have in mind a contingency plan that will let you recontact your caller rapidly. The agency moves quickly to fill the job order, so you must be prepared to answer the call rapidly.

If you do not think well on the telephone, you may find it helpful to keep you résumé worksheet at hand. Then, you will not find yourself groping to answer the questions. Also, keep a copy of your résumé handy. This is the document that the recruiter will be basing the conversation on so it is to your advantage to have it in front of you so that you are both working off the same piece, not just your recall. You should furnish the information requested and dig out your appointment book if the conversation turns to "when could you come in and talk to us further?" This is your cue that you are in consideration for a position. You will probably not be hired based on a phone call so you must be prepared to interview quickly.

The firm can most rapidly fill its job orders when it maintains a database of prescreened candidates. Filling a job order then becomes a matter of reviewing known individuals and determining their availability. As a new candidate, you must pass through the firm's screening process before you will be presented to the client for consideration. Once you are working for a firm and are on the database as one of its employees, the firm will monitor when your contract is due to expire and will consider fitting you into another contract. The first contract is the hardest to get. You are an unknown quantity. The firm must get to know you, your skills, and work habits.

If you have responded to a contract staffing firm's advertisement in a paper, the firm will be interested in when you will be available for work. Develop a clear picture in your own mind of your availability. A firm runs a newspaper ad for two purposes. First, it is looking to increase their pool of applicants with skills in high demand among their clients. Although it may not have an immediate opening, the skills are in demand, and the agency feels a need to maintain a pool of available talent on which to draw. Second, the agency may be looking for a special individual to fulfill a specific client's needs. There is almost always some urgency attached to the search even when the agency is restocking its pool of potential hires. If you can fill the bill, you will net a job.

When you are asked to come in for an interview at the contract staffing firm's office, you can expect that you will go through a standard job interview. The firm will be looking at you from two different angles. First, the recruiter will consider whether you have the qualifications, skills, and experience that their client needs. Second, the recruiter will be determining if you are an individual that the firm wants to hire. Reputable staffing firms know that their contract employees directly reflect on the firm and are careful to select potential employees who will reflect well on the firm's reputation. They also know the preferences and the culture of the firms that contract with them. They will consider not only your skills and qualifications but also how you will work with their client's organization. Before your interview, review the materials given in Chapter 2 and get ready for your new career.

Halfway There

If you make it through the firm's own review, your account executive (the recruiter handling the placement) will let you know what the next step is. Just because your firm contact seems impressed, you have only come halfway to your new job. You now must pass the client's review. To facilitate this, the firm usually submits a summary of your capabilities or even a copy of your résumé for consideration to its client. Although this process is standard, the exact paperwork that passes from firm to client may vary slightly. The recruiter and the hiring manager at the client firm will discuss your capabilities and your potential fit. The exact steps may vary depending on the client and the agency's relationship.

The relationship between the recruiter and the client is of utmost importance. A firm that has the trust of its clients and experience in providing contract employees that meet its needs will make placements more readily. This is one of the many reasons you will want to carefully choose the firm that you work for. A firm with a client base that knows, trusts, and routinely turns to the firm to meet its staffing needs will be able to keep its contract employees working.

If the you pass initial muster with the contract staffing firm and you are being presented for consideration, the firm will set up either a phone interview with its client or you will be asked to interview live with the potential contractor. While you are under consideration with a potential contractor, your references will be checked to ensure that there are no unpleasant surprises.

Applicants first coming into contract employment should not get lost in the mechanics of the process. It is very important to remember that the interview process is a two-way street. You are not the only person under scrutiny. Job hunters frequently forget in their eagerness to secure employment that they too should be evaluating the potential relationship. Contract employment requires that you develop these critical skills more highly. You need to evaluate not only the contract staffing agency but also the client and the job offered to you.

You should use this opportunity to review and assess how the firm handles you and whether the contract itself suits you. The hiring process will give a window into how the firm handles its employees. You will be working for the contract staffing firm. You should use the interview process to determine if this is the type of organization that you will want to work for. These considerations go beyond the usual pay and benefits that must be considered in taking any position. You should keep the following in mind:

Were you treated promptly and courteously by the account executive? Your account executive will more than likely become your key contact within the firm. If you and the recruiter do not hit it off, you may find yourself unable to communicate effectively with the person who will have the most impact on your career. Since you will not be working at the firm's site, you will need to quickly forge a strong communication link to your firm contact. Frequently, the recruiter is not just a screener who will forward you for further consideration in a typical corporate structure. The recruiter often becomes your career advocate.

Courtesy is a base measure of professionalism. If you are not treated with courtesy during this "courtship" preemployment phase, you can hardly expect to see a change once you have become an employee. If the recruiter will be your link between you, the contract staffing firm, and its client, you will want to determine if the person will be responsive to your needs when you are working remotely from the firm. In getting to know the firm, you will want to ascertain what sort of ongoing relationship you will be having with the recruiter: Will you be handed off to another person in the firm once you are hired? Will your initial contact remain your account executive for any length of time? This will help you understand the type of relationship you will expect. The account executive will probably not find it out of hand for you to make such inquiries as you move through the process. You may not get all of the answers on your telephone contact, but you should have a clear picture of the type of relationship you will have with the recruiter at least before you accept your first assignment.

Were your questions answered promptly and thoroughly? Your interview with the staffing firm is your opportunity to learn about the firm, its history, and other details that will impact on your acceptance of an employment offer. The account executive should be able to provide answers to informational questions and should be forthcoming with the answers. The firms that maintain the Internet sites given in the resource guide have made much of this information readily available on their web sites.

If the recruiter suggests that you are being considered for a specific contract, you should ask how long it will last. Use this as a springboard to ask how long the staffing firm's typical contract lasts. You can then determine if the contract you are being considered for is longer or shorter than most that the staffing firm handles. You should also use the interview to determine what types of employees the staffing firm handles. If you do not already know, be sure to ask: How many individuals does it employ? How long has it been in existence and in the area?

Did you feel comfortable with the answers that you received? To build a working relationship with the staffing firm and its recruiters, you will need to feel confident that you are being treated fairly. Your comfort level with the responses you get to your prehiring questions will color your long-term relationship. If you do not feel comfortable with the situation, consider whether it is the staffing firm and its personnel or the newness of the situation.

Did you like the companies the staffing firm places its employees with? You may like the potential contract you are being considered for, but if you cannot see yourself working at any of the firm's other accounts, you may want to consider your potential fit with the staffing agency. As a long-term employee of the firm, you can anticipate that you will move from account to account during your career with the firm. If the prospect of working at most of the employers the firm routinely supplies with contract engineers is unattractive, you will perhaps need to identify a different firm with a more personally attractive client list.

Can you picture yourself as an employee of this staffing firm? You will be working for the staffing firm, not for the business you are contracted to. If you find yourself excited about the prospect of working for the client or at the position you are interviewing for but have reservations about being associated with the staffing firm, you will most probably not find your employment satisfactory. Similarly, if you dislike the prospective contract but like the staffing firm, you will not be happy with your work during the contract. This may negatively impact your performance and your future with the firm and your career in contract employment. You must seek a measure of satisfaction with both partners in the employment arrangement—the firm and the client.

Do not hesitate to use your contacts and friends to help determine the reputation of the firm. If you know other engineers that work on a contract basis, you will want to ask their opinions of the firm. This is a time when it helps to have a network of business acquaintances on whom to call. You might even contact those using the firm's services to get a sense of how they are viewed. You will quickly learn whether the firm has an excellent or unsavory reputation. Your network of friends can help provide a fuller picture of the opportunity than you might get just by listening to the recruiter and client.

Can you picture opportunities for growth and career development within this firm? It is no secret that personal growth and satisfaction with one's career are powerful motivators for all individuals. Contract staffing firms as well as corporations are increasingly aware of the motivating power of a sense of the future. The downsizings and change in the world of employment have robbed many individuals of the excitement of expecting personal and career growth. Within the business community there is growing awareness of the despair that has crept into the job-seeking market. Today, businesses, eager to obtain talented employees, will try to alleviate this.

As you interview for a contract engineering position, you should be able to frame where it will fit in the continuum of your career. Jobs, as we have known them throughout the last century, are disappearing and morphing into new forms. The individual in this new paradigm has to become far more acute at evaluating the significance and importance of each career decision. A successful career in contracting is dependent on your ability to understand the changing labor market and fit yourself into a valued niche.

For a contract engineer each contract should contribute to your career objectives. Even if you are working primarily to put bread on the table, you need always to remember that we are each given just one irreplaceable life. It is our task to frame a unique and personally satisfying career in that lifetime.

Leaping through the Next Hoop: The Contracting Firm

In contract employment, you are part of a three-way employment triangle. You work directly for the staffing firm. The firm, however, has placed you with its client, whose needs it is seeking to satisfy. The firm has the difficult task of satisfying the client while keeping its own valuable employees working and earning revenues for the firm. As a prospective employee, it is your responsibility to make sure that the expectations of all parties are met. For yourself, you must make sure that you have meaningful, enjoyable, career-fulfilling work. You are key to the relationship and the satisfaction of all three parties. You will want to use the interview process to ascertain how each of the members of the employment triangle will work with one another.

When you are sent on an interview, you are given the opportunity to test these linkages and the potential fit between yourself and the client. Prior to employment, you will want to clarify the following:

Were the job expectations outlined by the recruiter the same as the person interviewing you? Lack of clarity of expectation is often a source of dissatisfaction in contract engineering. The client either poorly specifies the qualifications, tasks, and work that the individual will perform or somehow redefines them during the recruiting process without sharing the details with the recruiter. Later in this volume, we provide guidance for the manager who is writing the job order. The intent is to keep the manager from making the kind of mistakes that result in a less-than-satisfactory hiring or employment situation. The job order process and its fulfillment is often a test of the firm's relationship with the potential contractor. Where there is an excellent working relationship, there will be greater clarity and a smoother working relationship. Both parties will be more in tune with the needs of each other.

Not only should you seek specifics on the job and what tasks you will be expected to perform but it will be to your advantage to find out how the recruiter has represented you and your skills. Although recruiters are trying to fill a client's needs with the best candidate that is available, it is human nature to oversell the individual. You do not want to arrive on the contract with the contractor's having unrealistic expectations of your ability to perform on the job.

Get a clear picture of the working environment for contract employees at that organization. Since you will be working as a contract employee, you should probe during the interview the organization's posture toward contract employees. This is an area where you may find a range of varying outlooks. For example, as more companies grow increasingly dependent on contract employees, they are realizing that these employees are an important resource that should be treated with respect—not as disposable or instantly replaceable second-class employees. You will want to develop a clear picture of how you will be treated vis-à-vis the permanent direct-hire employees. Regrettably, there are still legions of employers that, although they are using more temporary and contract employees, treat them like second-class citizens.

Similarly, some organizations will claim that they treat contract employees *"just as if they were members of the team"* and then will severely limit the extent to which they participate. Many companies will limit the level of responsibility that they will assign to any nonpermanent employee; others look at the individual's skills and experience in assigning responsibility. It is not until you are actually working that you can test whether the firm is "walking the walk" and not just "talking the talk." Your professional network of contacts will often furnish a second, and valuable, point of view.

Identify who your immediate supervisor will be. When you interview for permanent employment, a critical component of the interview process is sizing up the potential working relationship between you and your prospective first-line supervisor. As a contract employee, you will want to meet and make this same determination. A positive side to being a contract employee is that if you find later that the person is not as compatible as you thought during the interview process, you will not have to wait for retirement or a promotion to get a new boss. When the contract ends, you will have the opportunity to move to another new boss. Just as you will find it helpful to ascertain the company's profile on contract employees, you will find it helpful to review the individual's perspective as well.

Examine the workplace, and picture yourself working there. The contemporary workplace is very unpredictable. You will want to know before you accept a contract whether you will fit in the organization. Before forwarding you for consideration, your firm recruiter will have already made a basic positive determination of your potential fit. It is now your turn to verify it. Minor issues can become major stumbling blocks to career happiness. For example, if you are a smoker, you might not want to accept a contract with a militant smoke-free organization. Many staffing firms will handle some of these details during the prescreening; however, take nothing for granted. It is your career, and in the new employment paradigm you are in charge of your career. Seize the opportunity to create your own happiness.

Examine the workplace, and picture yourself working in it. If you cannot see how you might fit, you will probably not find yourself very happy in it. If physical surroundings are important, don't think that just because it is a "contract with a time limit" you can expect yourself to adapt to a virtually alien work environment. While checking out the environment, consider your coworkers. Can you see yourself working with that group of people? Your ability to adapt to the total work situation will impact your performance.

Verify how you, the staffing firm, and the contractor fit together. You begin this process when you meet initially with the staffing firm's recruiter. The interview with the client will offer more information to help you understand how the three of you will work together. At the client interview, the client will be looking to see how you might fit and to determine if you can indeed fill the position. It is vital that you develop an understanding with your staffing firm recruiter and its client how employment details will be handled. Between the interviews with the recruiter and the client, you will want to learn the following:

Rules of conduct. What are the contractor's standard rules for employees. If you are a smoker, you will want to know where can you smoke. What is standard practice for lunch and breaks? What time would work begin or end? What is the policy on the use of telephones? Are there special safety rules and regulations? The answers to these questions are usually found in the contracting company's employee handbook. You will need to be familiar with these rules so that you may comply. Although you are a contract employee, you will be expected to live by the same rules that govern the contractor's other employees.

Confidentiality. Many technical contract employees have access to information that is sensitive in today's competitive technology-based global economy. Most staffing firms require their employees adhere to specific confidentiality agreements. Prior to accepting a contract employment position, you will want to know the agreements that will bind your work; take an opportunity to read them carefully.

Sickness and other missed time. No matter how conscientious you are, there is the inevitable bout with the flu, the unforeseen automotive breakdown that leaves you by the side of the road fuming about how you are going to get to work, or some other mishap. When you are a permanent employee, your channel for notification is simply to call your supervisor. As a contract employee, you will want to be clear on whom you should call and what the process entails. You do not necessarily want your recruiter to find out from the client that you are missing time that the staffing firm is unaware of. You obviously will want to inform your supervisor at the contract about your status immediately. The staffing firm will have a protocol for how to handle these situations. Don't wait until you are standing beside a flat tire to figure it out.

Plant closures, holidays, and labor actions. Part of your information gathering should include learning the policy and procedure for handling a weather-related plant closure. What is the staffing firm or contractor's expectations? How will your pay be handled? Strikes and other labor actions can present a difficult dilemma. If you are interviewing for and expect to work in an area that is likely to experience work slowdowns, strikes, and other labor actions, you will want to have clarification in advance of your employer's and the contractor's expectation of you in the event of such actions. If you will be expected to cross a picket line, you will want to know this in advance.

An orientation by the staffing firm and the contracting firm will sometimes provide answers to these and other questions that you might have. During your interview process, you should ask and determine what sort of an orientation you will receive. If your orientation is a map to the contracting firm's plant, a set of payroll paperwork, and a handshake, you had better ask some questions before you head off to your new job assignment. Since you will be working with the staffing firm and its client, you should also be sure to understand the specific role each plays in your career. The more you learn up front, the more likely you will be pleased with your working relationship.

It's All about Money and Such

Your references have checked out. The contractor wants you, and the recruiter makes you an offer. Now, you must decide whether contract employment and this contract in particular are in your future. With an offer at hand you must evaluate the economic issues and decide. The decision process at this point is similar to the one you already know from direct-hire work.

Although many of us are reluctant to admit it, we often accept positions based primarily on economic factors—pay, benefits, and such. Contract employees are no different. For many years, American workers have enjoyed a rising economy in which additional salary and enhanced benefits were the expectation and the norm. Today, wage growth has stalled with most businesses providing salary increases of roughly 4 percent per year.

Instead of enhanced benefits, employees are being asked to participate in paying for benefits that it was once the norm for the employer to pay. The growth of copayments on health care benefits is an example of this trend; so too are the 401(k) plans that have replaced traditional defined benefits retirement plans.

Contract employment, unlike many other industries, has seen a growth in the types and extent of the benefit programs offered to employees through the contract staffing firms. Those unfamiliar with contract employment inaccurately envision the contract employee as toiling on short assignments, underpaid, and provided with limited benefits and no future. Because contract staffing firms are in competition for highly skilled workers, they must offer competitive compensation packages to attract and retain these employees. There is some divergence though in the benefits packages offered. This is firm and market dependent.

As an engineer working on a contract basis for a reputable staffing firm, you will most probably find yourself offered a benefits package not unlike those of permanent employees for many companies in the area or industry. Benefits packages offered by staffing firms today often include a mix of retirement via 401(k) plans, hospital and medical benefits, dental plans, tuition reimbursement to enable you to further your education, and other options. You can also expect to earn paid holidays, vacation, and even personal days. Funeral leave and jury or military duty figure into the mix. Some firms, however, provide a much more limited offering for their employees. Others require lengthy periods of continuous employment before the contract employee is even eligible to participate in the programs. Since benefits are firm dependent, you will want to carefully evaluate this aspect of the contract mix.

When the package is roughly equivalent to a benefits package expected from another employer in your field, you will want to evaluate the package offered exactly as you might one offered by a traditional employer. Your benefits needs are a reflection of your life, and many packages are quite flexible, allowing you to tailor your package. No matter what type of package you are offered, you will want to consider these factors that are of special importance for contract workers:

What are the eligibility requirements? Although the staffing firm may provide benefits for its employees, there may very well be eligibility requirements that you must meet. These may require that you work for the staffing firm a prescribed length of time before you can sign up to participate in the medical or retirement plans. Where the requirements are service-time-based, you will want to weigh the eligibility requirements against the length of the prospective contract. This is crucial. You do not want to face a situation where the contract might end just prior to your vesting your benefits rights only to discover that any break between assignments might start the eligibility clock ticking again.

What are the provisions for employment breaks? A good contract worker can expect the staffing firm to work with him or her to ensure that there are no real breaks in employment. Breaks in employment are a reality that you may confront. You will want to know if the staffing firm makes some provision for employees dovetailing from one contract to another. Some firms provide "benching" provisions. These allow the contract employee a window of time between jobs without penalty or loss of pay or benefits. You do not want to have a break in your medical coverage or eligibility created by a brief interlude between contracts. You will want to be sure that you can maintain coverage.

What is the vesting requirements on profit-sharing or 401(k) plans? Many agencies provide 401(k) plans or include profit sharing. This permits employees to build up a retirement and share in the fruits of their labors. When the staffing firm prospers, so do the employees. You will want to check how long you must work for the firm before your profit-sharing portion of the 401(k) becomes portable. That is to say, how long you must work for the staffing firm to ensure that you do not lose the firm's profit contribution to your plan? Although you may very well like the firm and decide that you want to make a career with it, you will still find this information helpful for your long-term financial planning. The realities of the employment situation are that you might find it economically to your advantage to work on a contract with a different firm. You do not want to have your profit sharing become a set of golden handcuffs.

How does the firm handle situations in which your contract's plant is closed but your staffing firm is not on holiday? For example, the plant at which you are working goes on a 2-week shutdown to change over tooling for new models. You are not among the employees who will be working to make the changeover but will instead be on furlough. Will you be expected to file for unemployment compensation? How will this involuntary break in your contract be handled?

Does the compensation and benefits package meet your needs? Your satisfaction with your employment heavily depends on how you feel about your compensation and benefits. If you do not have the benefits that you need and a salary that meshes with your expectations, you will not be happy no matter what type of work situation you are in. It is, therefore, important that you evaluate your contract employment offer in the context of your total career, not just as a temporary way to earn a few bucks. In the next chapter, we will look at ways to build a career, not just get a job, in contract employment.

Summary of Key Points in This Chapter

1. Submit your résumé to a number of staffing firms at the same time to increase your likelihood of obtaining an interview and a job.

2. Select a staffing firm that provides you a range of possible employment opportunities.

3. In evaluating a staffing firm, remember that it will be your employer, and consider whether it is an employer that you can work for comfortably.

4. When interviewing with the staffing firm and the prospective contract, be sure to check to make sure that all parties have realistic expectations of you and your work.

5. Review any salary and benefits as you might those offered by a permanent employer; however, watch for eligibility requirements that mixed with your prospective contract could prevent you from obtaining the benefits that you expect and need.

4

Making Contingent Employment a Career

Jim, a midcareer mechanical engineer living in the Midwest, was laid off in 1992. Needing to ease immediate financial pressures, he took a job with a contract employment agency. Although he continued to look for a "permanent position" for 18 months, he never found a position that would offer the stability of his former employment. Some positions simply did not quite offer the variety of engineering challenges that he was discovering on his contracts. Slowly and somewhat reluctantly, he came to realize that employment, as he knew it during the early part of his career, might very well be a thing of the past. He has recently decided to abandon looking for permanent employment. He is still struggling with his decision. He continues to ask himself if he is wrong in accepting as permanent an employment model that he embraced originally as a temporary solution for meeting his financial needs.

When Jim first lost his job, his wife, Sarah, went back to work and has found her career both personally and economically satisfying. Their two children are now 11 and 17 years old. The son, the 17-year-old, has 1 more year of high school. Their 11-year-old daughter is in the sixth grade. Both Jim and his wife would like to relocate closer to their roots in the East. They realize that the ideal time to relocate would be within the next 2 to 3 years, before their daughter enters high school and while their son is in college. This would avoid undue disruption of the family's life.

As Jim and Sarah debate the possibilities of relocation, Jim once more finds himself asking, "Should I consider looking for a permanent position?" Given that he enjoys his contract work, he finds himself asking: "What are the opportunities for meaningful, satisfying contract work if we relocate?" He likes the regionally based contract employment firm that he is with and questions, "How will I identify and secure another contract employer that is as good as my current one?" and "Will another contract firm be able to keep me working as well as this one?" Jim needs to continue working and worries about applying for work at a distance. Neither Jim nor Sarah want to make a major move without jobs in hand. Their financial instability during 1992 and the impending college expenses for their son have left them very unwilling to accept any unnecessary financial risks. They are now looking for advice on what to do.

In the future three forces will shape our careers—technology, changing societal norms, and economic realities. As we examined in Chapter 1, economic realities and technology are already strong forces creating often unwanted and unpleasant change. These are strongly altering many of our societal norms. For example, they have already altered the fundamental employment contract that most American workers have known and held dear for generations. This implied contract, a holdover from the nineteenth century, was based on a shared set of values that traded employee loyalty and dedication to task for job security. The reductions in force of the 1980s and 1990s have shown that world competition and other economic forces can cause even the most stalwart companies like IBM to abandon these values.

The change has not come easily, and many engineers, as well as those in other professions, have unwillingly embraced the resulting new employment paradigm. In the new paradigm, the individual must take a far greater role in personal career planning and management and not expect a corporate involvement in what was once called "career development." While the individual's role expands, the corporation's diminishes. By reading this volume, you are taking steps toward interpreting the new world of engineering employment.

Today, you must address your own career management. You can no longer expect to find job security. You must seek to develop your own employment security. It is now your responsibility to make sure that you maintain the tools to stay employable.

You might ask, as Jim did above, "Am I kidding myself into accepting as permanent a solution previously only acceptable as a temporary solution to an employment crisis?" The evidence is stacking up that careers are altering their shape and path to accommodate the growth of the various forms of alternative employment. All indicators suggest that businesses will continue the trend toward downsizing and outsourcing. Department of Labor statistics show that even as the economy and the employment picture have improved, prosperous businesses are still shedding employees and restructuring. We are fooling ourselves to think that employment will return to the norms of the past. Those good old days are gone, and perhaps they were not so good after all.

By embracing the new reality of employment and determining a personal strategy for how to make it work to your personal benefit, you will be a pioneer. Like the pioneers of old, you may be able to lay claim to the choicest positions. As a pioneer, you will be among the first to know the new career geography. You will know how to successfully navigate in this new employment world when many of your peers are just discovering it. If all of us do not take charge of our own careers, we will be like riders on a roller coaster—on a suspenseful and scary short ride.

As the new career paradigm matures, those engineers who do not understand or embrace the new reality will find themselves out of time and place. They will be like those who failed to understand the need to develop computer skills. When the first waves of layoffs came in the 1980s and the early retirements piled up, it was the least computer literate who were made to understand that they were behind the power curve. It was often suggested that they really ought to consider a different direction for their career—early retirement.

Developing a career strategy that includes multiple employment options is a proactive approach that will insulate you from buffeting by continuing change. The need to develop a personal long-term strategy is most important when there are two careers in the family. The single wage earner was the societal norm just a generation ago. The wage earner was typically male and supported a nuclear family. Today, approximately 50 percent of women work outside the home, and there are almost as many singles living alone as married couples. Many of the singles are single-parent families, a group that has witnessed a tremendous growth. Even given the growth of single-parent families, the current trend of women entering and staying in the workforce translates into a very large number of two-career families. This has altered the landscape of career planning. Individuals must take a more proactive approach that will let them develop a career strategy that will yield satisfying work lives for all family members without undue hardship.

As both adults in the nuclear family have developed careers with increased responsibilities and financial remuneration, it is increasingly difficult for families to make a geographic move without suffering economic loss. With two wage earners, the economics are no longer clear-cut. Almost any move can easily translate into a win-lose situation for the family. Defining an individual as the breadwinner or one individual's career as less significant is now frequently not even a consideration. Many families today rely on both incomes to meet their expenses and provide security for the family's futures. Any career move that affects one member of the team will ripple to other members of the family.

The changing role of breadwinner has also created a number of changes in how careers are put together. Twenty-five years ago, it was the rare individual who turned down a corporate transfer. Today, companies are finding more individuals willing to risk the potential consequences of turning down transfer opportunities. The new corporate reality, where loyalty and longevity are no longer coins of value, is also making workers less willing to relocate. There is no assurance that a good soldier, willing to relocate for the good of the company, will be exempted from layoffs and downsizing. Many families do not want to risk moving the family into an unknown and equally unstable situation. They prefer to stay put and take whatever is dealt at their current location where their network of friends and social support are intact. To maintain forward economic movement, the two-career couple must plan their careers proactively as a couple whether the individual is working as a direct hire, on a contract, or other alternative basis.

Taking a Proactive Approach

A proactive approach to personal career planning requires that the individual (1) accept the realities of the changing workplace, (2) maintain marketable skills, (3) continuously watch employment trends, and (4) develop a self-directed career path with personal goals and measures of achievement.

Accepting the Realities of a Changing Workplace

It is one thing to read about the changes that have occurred in the employment paradigm. It is another to realize that it is not just something that will affect others. It is a reality for you. Those engineers with so-called permanent jobs should reconsider just how permanent their employment really is. Job security has actually slipped. One study by Princeton economist Henry Farber found that the typical male worker aged 45 to 54 had been with his existing employer almost 12 years in 1993, down from 13 years in 1983. This represents a significant alteration in employment longevity. Today, job insecurity is major morale issue in American business. It looms as the single largest issue in labor negotiations taking the place of workplace improvements and salary increases.

A personal assumption of loyalty and concomitant rewards may very well be misplaced. If you are currently working as a direct hire in a so-called permanent job, try this little test. To check the solidity of your employment relationship, place yourself in the position of a potential applicant for your own job. If you can clearly imagine your current employer hiring you and your skills for the salary you are making, you will probably not be declared surplus or redundant and subject to lay-off. You might ask, "How would I know?" Part of maintaining a self-directed career requires that you scan the marketplace and keep a regular check on your value. Just as you might check the worth of a stock portfolio, you should regularly check the value of your skills portfolio. Find out by checking ads, tracking the placements of others you have known through your professional network. It is important to keep an accurate pulse on your worth.

You should carefully consider whether someone with lesser skills or a smaller salary could perform all or part of your duties. Overpaid capacity quickly translates into losses for the company. Look closely at hiring patterns in your own company and others you are familiar with. Ask yourself honestly, "What specialized skills or knowledge do I bring to my job?" "Am I using them?" "Are they valued by the organization?" If the company is currently paying you based on skills that you are not using, there is little reason for it to continue paying a premium for an unwanted or unused resource. If your knowledge is readily transferable to a less costly employee or your skills are even somewhat out of date, you are vulnerable.

If you are overpriced for the job you perform, you become a target for downsizing. This may be through no fault of your own. You may have found yourself reconfigured by changes in the organization into a position that is too well compensated for its duties and responsibilities. Individuals in these positions should realize that at some point the organization will either redeploy the individual more effectively or sever the relationship. There are clear indicators:

1. If your organization has restructured and you are reconfigured into a project team that is somewhat short term, your career is in trouble.

2. If you have been placed in a job that is clearly not at the same, but a lower, skill level as your former job before the restructuring, you are vulnerable.

3. You have been told that you are eligible for no additional compensation. You are at the top of your salary range and have been passed for promotion more than once.

How do you know if you are overpriced? Consider what your package of skills and experience might bring if you were laid off. There are clear indicators of this as well.

1. If you think that you would probably have to accept a position that pays less than you are currently making, your salary may be a liability.

2. If there are readily available less expensive employees that could perform either at your level of productivity or higher, the company is overpaying for your services.

3. If your entire work group is overpaid or underproductive according to corporate or industry statistics, you and your work group might be potential candidates for outsourcing. Instead of reconfiguring or altering the staffing, some companies will simply rid themselves of the less productive unit by eliminating it and contracting its entire function to another more productive and less expensive provider.

4. If the work you perform is no longer viewed as essential to the business because of shifts in technology or other changes, unless the company has spelled out how it plans to use those with your skill mix, you are vulnerable.

5. If work done by your unit or at your plant is sent elsewhere, your group may be either overpriced or underperforming.

6. If there is rumor of work transferring from your site to another, check into it. If it is economically based, your unit or plant may be underperforming.

You are *most* vulnerable if you have a history of being a difficult worker. Although this is difficult to assess and requires careful self-examination, there are indicators.

1. If you have had a hard time getting along with coworkers or your bosses, you may not be perceived as a team player. In today's work environment, this is enough to derail your career.

2. If the only reason your employer would keep you is because of your longevity, you are not adding value to your employer. Your employer would be well served to be rid of you.

This may paint a harsh and cynical picture, but you must remember that your employer is in business to make a profit, not to run a social agency. Your employer must compete in an increasingly hostile and competitive environment and needs to have every employee add value. An employer has every reason to expect every employee to add to its competitive ability—that's what you were hired to do.

3. If have been passed over for a promotion, you may have plateaued and can no longer gain value. Unless you can clearly see retirement in the very near future, you need to be planning how you might respond to a change in your work arrangements.

4. The clearest indicator is, of course, if you do not get along with your boss or other key management personnel.

You must be perceived as an asset, not as a liability, in any work situation. Many long-term employees forget this relationship. They perceive themselves as irrevocably part of the organization. Their perception of themselves is completely linked to the organization. They assume that just as a family cannot fire the family members, the organization will react in the same manner. Many individuals rail against paternalistic organizations, but they want their own employer to treat their relationship with the same loyalty as a family. This is an easy mental trap, particularly if the organization prides itself on a "family" culture. Many laid-off professionals who grieve the loss of their jobs were blindsided by the layoff because their own assessment of their worth to the organization was unrealistic.

This is particularly significant if your organization has undergone a change such as a merger or a buyout. The culture may be undergoing a major change. A once-friendly family-oriented culture can change rapidly and reflect the culture of the new owners. A symptom of this is visible when you see members of your company passed over or forced out by the new team, or when all or a large number of your company's managers are replaced by those of the merger partner. This means the other company's culture is ascendant. If long-standing policies and managers are changing, so too may the corporate philosophy that supported them. Don't expect the company to stay the same when the managers at the top change and bring different values and culture with them.

A change in culture will often occur when a business shifts from a start-up owned by a few visionaries to a publicly traded, stock-issuing entity. The company becomes responsive to a different set of drivers. Whereas management once had to please only a few owners, when it is publicly held, its obligation shifts toward its stockholders. For those who have made their careers in start-up companies, particularly in high-tech firms, this can be a jarring reality that completely destabilizes the employee-employer relationship. The lament that the organization just isn't the same is indeed a reflection of reality, albeit unpleasant.

By developing a realistic picture of your career and the realities of your employment, you can begin planning for the future, not just awaiting it. You will find that this will reduce job stress and enhance your own mental and career health. If you find yourself lamenting about the changes that are occurring at your employment site, consider whether they are a signal of a profound change that you want to stay ahead of to protect your career.

Contract engineers and those who have accepted this reality can develop reality-based strategies that convert this negative picture into permanent employability. To do this, they must review their skills and knowledge base, consider how they might hone them, and then how they might convert this mix into a value-laden package. They are then able to offer this package not just to one beneficent employer, but to a number of potential employers. The challenge then becomes how to market their skills to obtain the greatest personal value. This approach puts them in charge of their career.

Psychologists suggest that having control over one's environment is a component of stress reduction. We frequently feel stressed by those things that create discomfort in our lives but that we cannot change. When we are caught in traffic and begin to feel stressed, it is because we are involved in an uncomfortable and unalterable situation over which we have little or no control.

Individuals who have created an employment skills package become like free agents in sports, able to go where they might bring the most benefit. A baseball team with a strong offense may find a utility player batting 195 superfluous to the team. A team laden with aging superstars might find the same player a perfect match to provide relief for their stars. The same holds in employment. This concept is based on a free-market approach to employment as opposed to employment based on a social contract.

Maintaining Marketable Skills

To stay competitive in the world market, employers are forced to improve productivity and quality continuously without increasing costs. Technology can provide only a measure of competitive advantage. The quality of the human capital at work in the organization—that which changes the processes, innovates, troubleshoots, and creates an environment that harnesses the capabilities of technology—provides the rest of the advantage. This means that every employee must be prepared to use the most current labor-saving or quality-enhancing workplace technology and know-how. There is no magic list of skills that you must maintain current because it varies by discipline.

There are, however, some types of skills that are advancing rapidly and require constant vigilance to prevent lapses. These include anything that relates to computerized technology. Industry's dependence on computers and technology continues to grow. It is incumbent on virtually every worker to stay current in the technology of their specific business area or risk being left behind.

An anticipated growth area is telecommunications. This is the first infrastructure improvement that developing nations undertake. With the tremendous improvements and moves made toward connectivity around the world, those with experience in telecommunications can expect worldwide employment to grow.

Similarly, materials and materials handling are changing as business discovers how to increase the productivity of our resources both renewable and nonrenewable. For example, recycling can be thought of as a materials utilization challenge. Environmental engineering and solid waste management are burgeoning fields. Landfill space is at a premium in most urban areas, and solid waste handling is a growing concern for industry and governmental agencies. With the cost of raw materials and their handling increasing as worldwide demand for scarce resources increases, the reuse of goods that were once considered waste will continue to provide economic opportunities and by extension employment. Water and sewage handling are integral to this.

The plants, equipment, and infrastructure that are the economic engine of the U.S. economy are aging. It is anticipated that there will be jobs in the future as new plants are developed to replace our aging industrial base. Engineers with experience and training will be in demand.

Government—federal, state, and local—continues to play an important role in regulating business. This means that it is important for technical personnel to maintain current knowledge of any regulations that might impact their chosen discipline. Again, this list is by no means exhaustive, it is given to jog your thinking. As you evaluate your skills and determine which you want to hone or advance, remember that you must consider whether you are advancing skills for which there are or will be employment opportunities. This is why it is important to continuously monitor your work environment.

Continuously Watching
Employment Trends

Every individual should, as part of a proactive career plan, evaluate the long-term viability and economic need for his or her specialty. In a dynamic society such as ours, demand for technical personnel with special skills will vary over time. For example, as the use of PCs and workstations has grown, taking over many functions formerly managed with mainframe computers, there is less demand for programmers who can write some of the languages such as Cobol and Fortran associated with older mainframe technology. Although there is still a need for programmers able to write in these languages, the need is not growing. A programmer looking to enhance his or her skills would find his or her time better spent learning a programming language that is growing in demand. Today, programmers with experience in C++, Unix, and the Internet are much in demand. The syntax for a new programming language may be different, but the logic skills are often transferable from one to another. Advancing skills and maintaining a current quiver of career-enhancing capabilities may require reliance on transferring some skills from an aging technology to an emerging technology. The difficulty comes when the individual does not or is unwilling to accept the need to transfer skills to another technology.

Several years ago many studio musicians faced a declining need for their skills. Synthesizers were readily replacing the live session musician. Many musicians were faced with either developing skills in electronic music, battling the competition for the slots available for live work, or rethinking their careers. The market had shifted. By keeping a watchful eye, engineers can protect themselves from the economic consequences of such shifts.

To protect yourself, you should go back to Chapter 2 (pgs. 28-29, Exhibit 2-3), review the list of skills that you developed for your résumé, then review the need for individuals with each of your specialized skills. Is the market growing? Check the want ads, and surf the employment sites on the Internet. Look for what is in demand. Perhaps demand is emerging for a skill closely related to one you already have. Then, you must ask yourself, "How much would it take to elevate my skills from one level to the next?" By taking a proactive approach, you can make the decisions yourself with time to make adjustments, as opposed to trying to play catch-up in a layoff or prelayoff situation.

Note that throughout this entire discussion on career planning, there is an emphasis on continuing education. Today, all individuals must look to maintaining long-term employability as opposed to long-term employment. If you keep yourself employable, you will find that employment follows. Organizations are configuring themselves to become "learning organizations," and to function in them, we must become lifetime learners.

You should also scan regional and area employment forecasts as part of your career planning process. These are routinely published in the newspaper and through various professional societies. Become familiar with and anticipate the employment cycles of your specialty and your geographic region. Although the living is quite good in some industries such as defense and automotive, it is, however, subject to employment cycles. If you choose to live in areas where a single industry dominates the landscape—Detroit, automotive; Seattle, aeronautical; southern California, defense—you must be ever more careful and understanding of these cycles.

A dip in the primary industry can create ripples of unemployment that reach throughout the region. Securing another job may not just be a function of maintaining your employment skills but may also hinge on the reality of available jobs. Don't wait until the demand softens and you are laid off to develop your contingency plan. By knowing the cycles of the industry, you can anticipate when you will need to change, before it is an absolute necessity. This forward and contingency planning is part of the development of a self-directed career plan.

Developing a Self-Directed Career Path

Traditional careers have followed a linear path. In this model individuals sought and secured employment with a single employer. Career growth and change were tied to its fortunes. In many instances, the employer offered a job with stability coupled with sufficient challenges to make going to work personally worthwhile. Many traditional individuals perceived their careers within the context of a single organization. They planned their long-term careers within a context of successive moves within their organizations. For them the promotion path was visible and logical and salary growth predictable.

Individuals only changed companies to seek faster promotions or because of financial or personal incentives. Some truly unfortunate, unpleasant, and occasionally incompetent individuals would be let go or lose their jobs for other reasons. This is why many individuals still have great difficulty separating the job lost through environmental changes from their own sense of self-worth. If you can develop an approach predicated on the assumption that instead of your losing a job that the company is losing you, you will find it easier to flex and adapt to change. You must in effect unhinge your self-worth from your identity with a single employer. This is easier said than done since loyalty is ingrained in us from childhood.

Career guides have long urged professionals to develop a career plan. In the past, the career plan was simply a thought-out version of the traditional linear path. A self-directed career follows a very different structure. Just as the Internet allows the surfer to access files from virtually any point, the self-directed career moves out of the traditional career box and accesses opportunities as they present themselves without regard for a straight-line structure. The path presumes a tremendous level of inherent instability in the environment and optimizes it. Change has a way of creating opportunity; the self-directed career planner scans for opportunity in a changing world and then takes advantage of it.

Out-of-the-box career planning presumes that the individual is in touch with the value of his or her skills and with career opportunities. To project and develop a self-directed career, you will need to have done the steps outlined above. You will also need to do some exercises in personal values clarification and develop a strategic focus for your career. When you seize control of your career and direct it yourself, it is a reflection of your own values and principles. To ensure your happiness, you will need to get in touch with them.

You will need also to develop a personal financial plan to accompany your career plan. In a traditional career, your employer bore much of the responsibility for your salary growth and retirement planning. Individuals would often target their personal retirement planning within the context of their firm's pension plan. Individuals who fail to look beyond the company for their long-term financial planning are those most impacted by the changes in their employer's retirement offering.

To stay competitive, employers are making changes to retirement health and other retiree welfare benefits. Corporate retirement plans are no longer immutable, and individuals need to recognize this in their planning process, whether they are working as a direct-hire or on a contract basis. This does not deny that an employer's plan should be a part of your overall plan, but you must be able to look beyond it—out of the box.

Many younger workers, for example, are convinced that the social security system will not be solvent when they retire and seek its benefits. They are looking for alternatives and are making their retirement calculations without its payments. All professionals would be well served to consider how to handle retirement planning in a volatile employment world. If your current employer has a 401(k) plan for retirement, how much of it can you roll over if you leave? At what point do you vest the company's contribution? If you are unaware of the portability of your current plan, you should investigate it as you develop your options.

Not only do you need to take a long view toward the obvious goal of retirement, you should also fit into your career plan how you will handle your other major financial decisions: buying a home, building and caring for the needs of your family, and meeting your personal needs. What about caring for your parents or for children who might return to a nest you once thought to be empty?

Every time there is a dip in a cyclical industry, the newspapers are filled with stories of home foreclosures and other economic misfortunes. Although there are instances when individuals cannot foresee personal financial disaster, there are many more times when a little realistic planning could have prevented a disaster. This volume is not intended to serve as a financial planning guide, but we urge you to add financial planning to your career plan.

Most individuals work for personal satisfaction and to meet financial goals; therefore, the two must work hand in hand. Financial planners will advise you to set specific goals and build timetables for making them a reality. Your career plan should also reflect this same type of methodology. By setting clear career objectives, you will find that your path is easier to define. You should write your goals down and share them with family members who may be instrumental in helping you achieve them. Family members will often add another dimension to your planning. Your career plan links directly to their happiness. In some families there is a recognition of the changed world of work by all members from senior citizen to high school age students. In others there is an unwillingness to accept the reality of change. If you make change without ensuring that the other members of the family share your view of reality, you may find that they do not provide the encouragement and mutual commitment to your career that you need.

For many individuals, career goals are primarily financially driven. If this is the case, you will want to direct your career to meeting them. This will mean selecting employers and employment opportunities that are specifically directed at increasing your compensation. Some individuals with strong financially driven career paths have found the solution to meeting their goals fastest rests in expatriate work. By working abroad for a few years, they are able to rapidly accumulate the money they want to realize their goals.

For other individuals this option is not even within the scope of possibility. They are geographically anchored. Just as some individuals will roam for economic gain, others will trade economic gain for an opportunity to live and work where they grew up. As you clarify your career goals, you will find yourself also developing a picture of what drives you. To help clarify your goals, it is sometimes useful to consider the most aggressive strategy you might use to move toward the goal most rapidly. If you discover that the strategy is either too drastic or unattractive, you may want to revisit your goal and identify the other influences. You may not have found the central values that govern your life.

Many individuals, however, set career goals that help meet specific family needs. For example, Matt, in the story that started this chapter, has a strong desire to consider the needs of his wife and children in planning his career moves. Although he wants to return to the East, he is not willing to strike out immediately. Caring for his family's needs makes him willing to adapt the timing and how he will meet this goal.

As you go about setting your goals, do not frame them in the context of your current employment. This will trap you in linear career planning. You will stay in your career planning box. State your goals without language that includes reference to your current employer. This method does not suggest totally tossing your employer out of the equation, but rather moves it from a central to a supporting role.

Your second step is to develop a series of broad strategies that will lead to the fulfillment of your goals. All of your goals should be supported with a strategy. You cannot achieve a goal without a strategy to move you toward it. Again, do not shape your personal strategies in terms of your current employer. You will find it helpful to write out the strategies and see if they flow logically from your goals. If you select and then follow a strategy that does not relate to your goals, you will not develop a working plan that will lead to you meeting them. In support of a personal financial goal, your strategies might include developing new areas of expertise, changing skill areas or industry, or moving across the country.

Your last step is to develop the specifics of how you will make your goals a reality. These are your tactics. They may include taking a course within the next year that will lead to developing the new skill that you have targeted in your strategies as essential to moving toward your long-term goals. Just as you need strategies to support your goals, you may find that each strategy will require multiple tactics to achieve the results you want. You may find that many of your tactics are in essence short-term strategies. Don't worry if this is the case. Your biggest concern is that your strategies and tactics in the end support your career or financial goals.

Even the best career plan will need to be revisited on a regular basis. First, you will want to determine your progress toward your goals. You will also want to reevaluate whether you need to make adjustments in the tactics you have selected. That you have laid out a plan does not mean that you should follow it blindly. The world of employment is so volatile that you will want to make sure that your strategies and tactics are valid at least every 2 to 3 years. This also gives you an opportunity to reassess changes in the broader environment. For example, just a few short years ago, the Internet was a tool used by researchers only; now it is a part of the American business landscape rapidly spinning off job opportunities for those who perceived its value and prepared themselves to take advantage of it.

Individuals familiar with strategic planning will recognize the methodology given above. It is simply a standard strategic planning methodology applied to career planning. The question still remains, how does this methodology support an out-of-the-box, self-directed career plan? By unhitching your planning process from a dependence on your current employer, you open yourself to exploring new opportunities. By developing strategies that relate to your long-term career outside of your current job, you will be stepping outside of the comfort zone of your current employment. The real move from linear career planning rests in the type of tactics you develop and those you actually use.

Nontraditional Employment Tactics

When you take your current employer out of the center of your career plan, you open your options. As you develop your tactics, you should reconsider those elements of your job that you find most rewarding. You will enjoy the greatest fulfillment if you develop tactics that move you into areas that you find most appealing.

Many engineers are drawn to engineering because it presents so many interesting problems. When engineers are surveyed about what they enjoy about their chosen profession, they will usually reply that they enjoy solving problems. They are happiest when they are solving complex technical problems. They want variety and do not enjoy jobs where they cannot call upon their technical problem-solving skills.

Contract employment with its short-term, project-based approach is a career tactic that these individuals should explore. Contract work is inherently problem-based. The individual is brought in to solve a problem or complete a specific project. The contract usually ends when the problem is solved or the project completed. The engineer then moves on to a new project with a new set of problems just waiting to be solved.

Technical skills provide the individual with career power. Expertise is the key to many fulfilling opportunities. As you look at tactics, do appraise the value of your unique blend of skills and experience. By detaching your career plan from your current employment, you should see more clearly if your skills are actually bringing you their economic value. Your tactics may require that you exploit your skills more readily.

Again, many individuals are drawn to contract employment because it allows them to earn what their skills are worth independent of other factors. They become highly valued, well-compensated hired technical guns. Since corporations have moved to increasingly lean staffs, they are no longer willing to stockpile valuable skilled individuals at compensation levels reflecting their worth. They either let the employee go or underpay and underemploy an individual willing to accept the situation. These underemployed individuals find themselves in jobs that do not fully use their skills. If your skill review suggests that you are underemployed, you may find that you can earn more nearly your worth as a proverbial free agent—a contract employee.

If part of your career plan is to develop experience in specific areas, you will also find that contract employment will provide experience that you can transport from one job to another. Most contract employees will admit that they learn something new on every contract. This is part of the value of a contract employee. They bring a new perspective to an organization, but they are also equally likely to learn something new that they can take to their next contract. Early career individuals will find this tactic most useful in gaining the skills and experience that will allow them to move forward. As contract employees, they should focus on taking assignments that will move them toward the experience level that they desire for meeting their next set of career goals.

If your personal plan includes changing geographical locations, you will find that contract employment can provide you the opportunity to move without changing employers. If your move is within a geographic area served by your contract staffing firm, you can more easily move than you might if you were tied to an employer that cannot place you in a job near where you want to be located. There are a number of contract staffing firms that cover large geographic areas. When you are an employee of one of these agencies, you have increased mobility.

If you are just starting on your career or are very cautious about the culture of an organization that you will expect to spend several years with, you may find contract work an excellent choice. Many companies today are equally concerned about how an employee will fit in their culture and choose to try the employee through a contract firm before extending a permanent offer. Temp-to-perm employment lets the organization get to know the individual and his or her work habits. This also reduces its recruiting costs since the staffing firm assumes these responsibilities.

You may find as you build your out-of-the-box career that you combine contract or alternative employment with stints as a permanent, direct-hire employee. You will find that by choosing an out-of-the-box career path which focuses on seizing opportunities that provide means to achieving personal goals, you will become more in charge of your career. Your control over this part of your life will reduce anxiety and allow you to enjoy career fulfillment in times of upheaval in employment. By understanding what you like the best about engineering and reassessing how you can mesh your work delights with making a living, you can bring positive energy into your career and create personal happiness.

What about Being an Independent Contractor or Consultant?

Many individuals find themselves chafing at being an employee and yearning for the flexibility of working for themselves. There are numerous advantages to working for yourself; however, there are equally numerous drawbacks. When you work for yourself, there is a sense of being in control of your own destiny. There is also the fulfillment that comes from being an entrepreneur. As a consultant or contractor, you have the freedom to select the firm and location that you want to work for. You, not an agency, get to select the projects that you will work on. This can provide an enhanced self-worth and personal feeling of reward for your work. As a consultant, you also get to negotiate the terms of your employment and work as your own boss. If you are at heart an entrepreneur, you can develop your own business.

The role of business owner is not suited for every individual. Unless you are a risk taker and comfortable with the financial insecurity inherent in developing a business, you should carefully consider the positive and negative aspects of this work before embarking on this course. The risks and rewards are the same as small-business ownership. With a consulting business, you can anticipate start-up costs as with any other business and considerable income instability as you develop your business. Since you will be in business for yourself, you will be subject to the same regulations and paperwork as any other business. Whether you have one employee, yourself, or several associates, you can still expect to have to work as an entrepreneur—a small-business owner.

As a business owner, you will be responsible for the marketing that keeps the revenue stream of your business constant. If you work as an independent contractor for a single firm, you are always in jeopardy of losing your sole source of support. You will in fact have all of your eggs in one basket. Try though you might, as a contractor, you will not find be as privy to the internal grapevine as a permanent employee. This means that organizational changes can catch you unaware.

Further, although you will enjoy a measure of independence, you will have all of the paperwork of being an employer and no tax benefits. The IRS takes a dim view of independent contractors and provides no tax incentives. Because of problems with misclassification of employees, some companies are reluctant to work with individuals as independent contractors. They simply do not want to risk problems with the IRS.

To protect your earnings, you will have to market your services to multiple organizations. This is time-consuming. If your personality is ill-suited to this type of activity, you will find yourself dreading calling on new customers. This will undermine your earning potential. As an independent business owner, you answer only to yourself, and no one else prompts you to make those necessary but dreaded marketing calls. Many independent consultants find that they must spend up to 50 percent of their time marketing. If you are a procrastinator, you will find that you have no one prompting you not to waste time. You must be an aggressive self-starter to succeed as a consultant.

If your technical skills are your most valuable asset, you do not want to do anything that will take away the opportunity to earn from them. This is particularly important if there is an imbalance between your marketing and technical proficiency. Every hour that you spend marketing is less productive than a similar hour spent doing the work that you are best at. Many individuals find that they must work many hours to compensate for this discrepancy or accept decreased earnings. You should consider that the total number of hours available per year for work is approximately 2,000. If you must spend up to 1,000 handling administration and marketing for your fledgling firm, you can quickly see how much you need to earn from the other 1,000 hours to make your current income. Research has shown that consulting and independent contracting do not necessarily yield greater financial reward for participants. If the rewards are not there, the question becomes, "Why accept the risk?"

As an employee of a contract staffing firm, the firm assumes these marketing responsibilities. Since the firm employs individuals with specialized sales and marketing skills to handle these functions, you enjoy the advantage of having trained talent working for you. When you work for a contract staffing agency, you can avoid much of the paperwork associated with working as an independent. The agency provides employer-related functions such as ensuring that your payroll taxes and withholding are paid. The firm also handles FICA contributions, pays unemployment taxes, and workers' compensation. The firm provides the paperwork support that many small businesses consider onerous. This clears the way for the individual to concentrate on using the technical skills to best earning advantage.

If you put off filing paperwork now, just consider the impact of failing to file a quarterly estimated income tax with the IRS. Some independents find that they have to hire accounting services to assist in keeping their books and making sure that they do not run afoul of the taxing bodies. If you run your business from your home, you will have to consider the potential for zoning violations and local regulations.

As part of the start-up of your consulting business, you must also plan to provide the technical support required for today's home office. This includes at the minimum phone, fax, answering machine or service, computer support for billing and proposal generation, and a designated office area within your home. Setting up your consulting business will need to make an investment in the equipment and support that your clients demand and expect. As an employee of a contract staffing firm, you have the support of their offices and do not have to make any of these outlays. Furthermore, you will probably have access to its benefits program. The firm is often able to get better insurance rates and frequently provides profit-sharing retirement plans.

As an independent contractor, you would have to secure your own insurance and provide all of the support for your retirement. In addition, any retraining or skills enhancement would have to be undertaken at your own cost and on your own time. When you are working as an independent contractor, time spent retraining costs double. You not only pay for the training but you lose time that directly translates into earnings. Many independent contractors do not advance their skills simply because they cannot afford to lose the time from their businesses to upgrade their skills.

Independent contracting and consulting are not suited for every individual at every stage in their career. For example, an engineer early in his or her career may not have the skills and experience sufficient to successfully win consulting jobs. These individuals might also find that their network of contacts is still too small to rely on as a marketing source. By the time they are in their midthirties, they should have sufficient experience and marketable skills to branch out on their own. An individual further along in a traditional career may have too much invested in a traditional career path to risk the financial dislocation of starting a business. Small businesses are usually best begun by those with the energy of youth tempered with enough experience to make them knowledgeable about their chosen field.

Those ideally suited for consulting or independent contracting in engineering are usually individuals near the end of their careers or in early retirement. With a secure vested retirement and limited career aspirations, a stint as an independent contractor or consultant to a former employer may be just the solution for finishing a career and easing into retirement. This option is particularly attractive to those individuals who accepted juicy retirement packages that included paid health care until they achieve Medicare eligibility. For these individuals, consulting work simply fills a gap in their earnings. They do not look forward to having to rely on the growth of their consulting business to provide income during peak earning years. Unless you are at an appropriate juncture in your career, consulting and independent contracting can be headache filled.

Summary of Key Points in This Chapter

1. Technology, economic reality, and changing societal norms are altering the employment reality and how careers are put together.

2. Coping with the changing career landscape requires accepting the realities, maintaining marketable skills, watching employment trends, and developing a self-directed, out-of-the-box career path.

3. To develop a self-directed career, you should clarify your goals and personal values, develop strategies consistent with these goals, and implement tactics supporting the strategies.

4. Out-of-the-box career planning may include periods of employment in a variety of career modalities—contract and permanent.

5. Reserve working as an independent contractor or consultant to a former employer as an end-of-career strategy.

5

When Are Contingent Employees the Answer?

Jack and Morris, two engineering managers with a southern manufacturing plant owned by a large multinational company, were discussing their departments over lunch. Jack, the more senior manager, has recently been given responsibility for a growing department. He comments to Morris: "It all used to be so easy. When the workload increased, I could always justify hiring more staff. Now, not only do I have to justify the hire but management still doesn't want me to hire the staff I need. They seem to think that I should just get the people through a contract staffing firm."

Jack expected to receive a little sympathy from Morris. Instead he was quite surprised when Morris answered: "What's so strange about that? I have been using contract employees for several years now. They give me an excellent opportunity to manage my payroll costs and still get the work done. Management sure likes what it does to the bottom line. What's so tricky about hiring contract employees? I understand that there is some discussion at corporate about outsourcing all of the design work done at the Midwest plant. So, if you don't like using contract

*employees, you just wait." The two men continued their
lunch, but Jack was still troubled over how he would staff
his department. Why was he being pressed into using
contract help? Could contract employees really fulfill his
needs? What about this outsourcing rumor?*

Increasingly companies are choosing to use contract and other contingent staffing options to help manage their human resources costs. Today, organizations can choose any number of flexible staffing options that include contract workers, leased staff, part-time employees, temporary employees and self-employed contractors. Organizations that need highly trained technical staff are frequently choosing to use contract staff as an alternative to permanent staff, not just as a temporary replacement for existing staff. As we examined in Chapter 1 (pgs. 4-9), contract staffing provides management many benefits.

The key advantage, and the reason many companies use contract staff, is that it lets them stabilize their workforce and reduce personnel costs. By using contract staff to meet peak needs, businesses can eliminate some cyclical contractions and expansions that have previously resulted in inflated personnel costs. When a business can readily flex its staff to meet demand, it is also able to provide a more stable work environment for its permanent staff. The constant threat of "who will the next layoff affect" is eliminated.

In a downsized corporate environment, human resource departments are smaller than in the past. When employees are contract staff, many of the human resource administration functions are transferred to the staffing firm. It is easier to pay an invoice for multiple contract employees than to complete all of the payroll paperwork attendant with having these same individuals on the payroll. Benefits administration represents a major headache for most companies. With a contingent workforce, this headache is the staffing firm's.

Since the contract staffing firm is responsible for recruiting the highly skilled staff needed to meet a contractor's needs, the individual firm can also reduce its recruitment costs. When the economy is humming at full employment or in proximity to it, skilled employees are in great demand. Securing candidates and hiring staff can prove a difficult, time-consuming, and costly process. The contract staffing firm, on the other hand, is by definition expected to keep a pool of employees or résumés from which to recruit. Because firms providing contract technical staff are specialists in technical staffing, they often draw on a broad network of potential hires that would be impractical and too expensive for a corporation to maintain. The ability to transfer the difficulties of recruiting highly skilled employees who are in short supply is often a driver for companies to use contract employees during times of full employment.

With historical staffing modes, a company first sustained recruitment and hiring costs as its employment needs increased, and it added more staff to meet growth. Then, it continued to sustain costs while orienting and integrating the hire into the team. These costs were sometimes hidden in decreased productivity as the new hire supposedly "came up to speed." The length of time needed for this process varied depending on the job and the individual hired.

Because managers were hiring a supposed long-term position, the hiring decision dictated that they weigh how the employee would possibly fit over the long term as well as meet the short-term need that triggered the hire. This sometimes resulted in short-term compromises that would negatively impact productivity. For example, the firm might hire a recent graduate, knowing the limitations of his or her experience, and then expect to train and shape the individual to meet the company's long-term needs. Today, the competitive nature of business requires that all employees have immediate application at close to full productivity. The cost of the learning curve is no longer an acceptable cost of doing business. The contract employee is expected to work at a high level of productivity almost immediately and with the exact skills needed for the task at hand. When a company builds a working alliance with a staffing firm, it may even transfer the costs and the administration of new employee training to the staffing firm. In Chapter 8, we will discuss how to build a strategic alliance with a staffing firm. In these relationships, the staffing firm often provides a liaison for their contract staff to use as a contact to provide the human resource support necessary to manage the contract employees.

Technology is moving so quickly today that skills are developing the perishability of food products. Most companies recognize that they must continue to train their staff or expect an employee's valuable skills of today to erode. The cost of this training adds up. When a firm chooses to use contract employees, it can often avoid some of these training costs, particularly when they impact either a very small number of employees or involve a very complex technical skill set. Contract staffing firms are aware of the need to maintain the currency of their employees' skills and will often provide employees the ongoing training as part of their business. The overall result is that many firms have found that with contract employees they can substantially reduce recruiting, hiring, and training costs.

When Are Contract Staff the Answer?

Many companies through experience have identified specific situations in which contract employees are the answer. Others have developed sophisticated cost analyses for evaluating their staffing scenarios. These models will usually contrast recruitment, wages, benefits, productivity, and training costs for using permanent versus contract or other contingent staff. Any manager considering using contingent or contract employees should consider applicable cost recovery of these items. The hardest to measure are productivity and training costs. A permanent staff worker familiar with all of the corporate infrastructure and the task will achieve greater productivity than a new employee. Even when the contract employee arrives with appropriate skill training, there is some inevitable loss of productivity associated with the settling-in process of bringing on any new employee.

No matter what costing scenario is used (none is offered here because the use and interpretation of these basic variables will differ widely depending on the situation), most organizations have found that contract staff are best-suited for time-bound projects. This allows the company to end its commitment to the employee with the completion of the project. There is no need to retain or retrain the employees hired for the project. In the past, many companies would seek other opportunities to deploy the project's staff rather than simply let them go. This sometimes resulted in expensive underutilization of staff resources or in increased personnel costs from downsizing.

A completely flexible workforce is a double-edged sword. Although it is easy to end the commitment to the contract employee when the project ends, many projects have elements that touch on the organization long after they are completed. An organization using contract staff for project-based work must make sure that it provides some means to retain the necessary corporate memory that will allow for continued positive impacts from the project, if necessary.

When there is a need to ensure continuity between the project's work and the ongoing business, firms have found that they can achieve this by combining permanent and contract employees on the same project. The permanent staff are often in management roles and are specifically charged with carrying the project into its next level or phase. The permanent staff will often have more general, managerial skills; whereas, the contract employees are highly skilled specialists, hired to handle specific technically demanding aspects of the work.

Managers trying to build this type of organization should be cautious in placing contract employees and permanent staff on a project where they both perform identical tasks. This creates tension and can create legal entanglements. When the work is project-based and the permanent employees on the team are subject to reassignment, it is defensible to have the contract and permanent employees work at the same tasks. It is also defensible when it is clearly stated that the need for the contract worker is driven by workload demand. When permanent and contingent employees work together at the same tasks, it is difficult to provide the necessary separation between the two workforces. Motivating and rewarding permanent employees for doing the same task as contract employees not subject to similar rewards pose additional and avoidable managerial problems.

The ability to combine short-term technical expertise without jeopardizing continuity allows firms to reduce their financial risk inherent in developing new products or designs. They can readily adjust the size and scope of a project and take far greater strategic risks without suffering far-ranging consequences resulting from overstaffing. Organizations that use contract staff often are aggressive project managers.

Today, with the emphasis on corporate culture and proactive management of diversity in the workplace, many firms are more carefully screening prospective employees. Some companies have chosen to use contract hiring as a prescreening strategy for prospective permanent employees. By hiring an individual on a contract basis, the firm can screen the prospect in real time and prevent some costly hiring mistakes. Firms can easily evaluate work habits, training, and cultural fit without making a permanent commitment to an individual. Since the contract staffing firm does an initial screening before sending the prospect to the contractor, a job candidate is in effect screened twice—first by the contract staffing firm and then by the employer. This extra layer of professional screening provides another risk-reducing checkpoint.

It's Not without Drawbacks

Although contract employees may seem to present an ideal solution for flexible staffing and even recruiting, there are drawbacks to using contract employees. It is a strategy that does require careful consideration. The ability to manage contract staff cost-effectively directly relates to an organization's ability to manage its projects. If the firm or department's management staff has difficulty managing projects (keeping them on time, within budget and scope), the same inefficiencies will creep into the project no matter what staffing strategy the organization uses. The use of contract employees can force the manager to consider more carefully why an activity is needed and what value is added by each task the employee performs, but it cannot make the manager more skilled.

Because the contract employee is a hired outsider, not just a redeployed department member, the manager is often forced to more carefully evaluate the hiring decision and the work that drives it. Managers who are unable to reflect on this or who are poor at determining staffing needs can actually add costs to their projects with contract employees.

Some managers find it difficult to bring new people into their departments. They are either so involved in solving pressing problems—the proverbial firefighter type manager—or they lack the perspective to understand what orientation new employees will need. In the next chapter, we will give some advice for how to rapidly integrate the contract employee into the workplace. Any manager choosing to use contract employees should first consider not only the project and the tasks that are to be accomplished but should also look inside himself or herself to determine if this type of employee suits the manager's style. It may signal a developmental need for the manager inasmuch as flexible staffing and contract staffing in particular are growing in popularity as a solution to manpower deployment.

Managing a workforce composed of permanent and contract staff also requires managerial skill because managers must be able to blend a group of employees with often divergent outlooks and perspectives than a staff composed entirely of permanent personnel. Managers must make every effort to integrate the contract staff into the flow of the work; otherwise, they add potential inefficiencies that can actually overmatch the benefits of having a flexible workforce.

If the managers cannot create a seamless work environment, they will end up with more than just two separate workforces, one composed of permanent staff and the contract staff. They can expect to find themselves overseeing a hostile environment filled with dissension and other human resource problems.

The selection of appropriate situations for using contract staff is key to building a seamless—virtual—organization. Since project or peak load work best suits contract staff, it is incumbent on managers to determine what portion of a project or even their department's workload can be separated into discrete projects. If they can create a series of short-term projects, in effect, carve Swiss cheese holes in the work, then they can more easily add contract staff. The key is to make sure that the project work has clear, discrete end points.

Through careful planning, managers using contract employees can save on new-employee orientation costs. These costs and lost productivity during corporate specific training must be part of a cost-benefit decision of using contract staff. To minimize orientation costs, managers should designate projects for contract employees that initially require a minimum amount of corporate-specific or proprietary information. This allows the manager to minimize orientation costs. Putting a contract employee in a situation that requires extensive knowledge of the company and its systems will undermine the employee's productivity. This can create inefficiencies and sabotage the cost-effectiveness of the contract employee as well as the company's satisfaction with the individual's performance.

Contract employees are inappropriate for technical jobs that require extensive boundary spanning. These jobs require the individual to interact with many individuals in other nonrelated areas of the company. Sales personnel are classic boundary spanners. They must relate with accounting, distribution, and manufacturing as well as the customers external to the company. The time it takes for a new salesperson to learn to thread the corporate infrastructure is often a significant limiting factor in initial sales performance. Ongoing sales training usually focuses in part on reducing the process loss attributable to the salesperson's boundary-spanning role. Some companies with products that do not require extensive technological expertise or have simplified fulfillment are using contract employees for their sales force. Their use, however, requires a simplification of the processes and the product to ensure maximum productivity. However, for highly complex products with long sales cycles, the contract employee is usually not an optimal solution to sales staffing woes.

In engineering, not only will the contract employee assigned tasks with extensive boundary spanning find it difficult initially to complete them satisfactorily, but he or she will also lose time while coming up to speed with the organization's systems and structural framework. Many contract employees are quick studies because of the requirements of their unique careers; however, it is unrealistic to expect them to perform without the corporate and other proprietary knowledge required for a project. If it is possible to reduce the amount of specific corporate information needed and the degree of boundary spanning required, you can reduce the training curve for your contract employees and increase their immediate productivity.

Choosing the right situation is only the first step for successfully employing contract staff. The second step is to develop a specific inventory of the skills that the contract worker needs. The market value and availability of the skill inventory both internally and in the area's employment market are key variables in the cost-benefit analysis.

Managers must learn to develop accurate skill inventories. Managers who typically have difficulties developing job descriptions that mesh with their employee's actual duties will find that they are often unable to employ contract employees who will meet their needs. To use contract staff successfully, the manager must have a much more accurate grip on the real requirements of the job than ever before. The contract staffing firm will endeavor to find an employee with the specific set of skills the manager purports to need, but if the hiring manager gives the staffing firm wrong or incomplete information, the candidates presented and hired will not satisfy the staffing needs.

To get extra help to guide your firm in obtaining the promised cost savings, choose a contract staffing firm that goes beyond providing staff as just a commodity. The staffing firm should not only understand your organization's technical needs but also its culture and values. If it does not, your firm may find that using contract staff is similar to having a permanently revolving door with no limit to the number of people who come and go through it.

If hiring freezes and headcount-driven internal requirements force you to use contract employees, you may be confronted with having to realign existing staff before making a decision about hiring contract employees. Some companies have frozen permanent hiring but have left managers flexibility in using contract employees. This requires that the manager reconsider not the slot that is needed but rather the workload of the entire department. If the change in demand has come as a result of a retirement or employee movement, it is tempting to think of the position in terms of the individual who most recently filled it. The manager should rather use the opportunity to redeploy staff as needed to ensure that permanent employees are assigned to core tasks while the contingent staff handles those tasks more likely to be of a temporary nature.

Outsourcing and Other Options

Some companies such as Morris's have chosen to outsource entire functions or departments. By transferring the management of specific functions to an outside vendor, the company frees up valuable managerial resources to focus on the company's core business. Many companies, severely threatened from outside competitive pressures, have found that by outsourcing nonessential functions, the organization can focus its entire resources on meeting these external challenges.

Some companies that dramatically reduced their headcount in the last 5 years now are finding that they are suffering from a managerial and technical anemia—they have too little support to grow with. They are now too lean to handle new initiatives. When they stripped out the middle-management layer, they removed the slack that allows for growth.

Children's shoes are bought with a little extra room in the toes to allow the child to grow without growing out of the shoes prematurely. The same is true in organizations; there must be enough capacity in the system to allow managers to assume additional duties associated with new projects and initiatives. There must also be sufficient talent waiting in the wings or readily available to fill managerial positions vacated by promotions and growth. During times of full or near-full employment, it is difficult to find capable individuals without having to pay a premium price. This is a tradeoff that companies today are increasingly faced with. One solution they have found is to decrease the load on managers of tasks and responsibilities that do not relate to the core business. To increase management's capacity to handle new initiatives, some organizations have found the best solution is to concentrate on the core business and outsource other management-sapping nonessential areas.

The best targets for outsourcing are functions least related to production such as janitorial services, data processing, and equipment or building maintenance. These are discrete functions that an external company can readily do with limited risk of disruption to the firm's core business. As companies have grown more comfortable with this strategy, they have expanded their use of this strategy to include other areas such as legal services and distribution. It can be expected that as more firms successfully try this strategy, those servicing these firms will become more agile in providing top-quality services to meet their clients' needs.

The companies using outsourcing have found that they can more easily track the costs associated with the specific functions. The cost-benefit analysis becomes very straightforward. When the company outsources a function, the outsource contract represents a specific dollar amount that will go toward funding the entire function. For example, the contractual cost of an outsourced maintenance contract becomes in fact the cost of maintaining the equipment. Some managers contend that there are fewer hidden costs, and that outsourcing reduces the need for administrative overhead. This is a tempting oversimplification.

Some companies have chosen to outsource highly technical areas such as data processing, information technology (IT), and design engineering to ensure an adequate supply of skilled technicians. Frequently, they have adopted this approach to reduce their recruiting, hiring, and training costs. In areas where highly skilled labor is scarce or at a premium, the savings can be significant. This also transfers to the outsourced contractor responsibility for maintaining a steady supply of labor. The underlying assumption is that the contractor will have a recruiting network beyond the company's range and thus meet the ongoing need. However, it is tempting to think that just because the responsibility is transferred to another company, that it is somehow immune to the recruitment problems presented by a full-employment economy. The outsourcing firm, although specialists in providing the service, will sometimes be forced to pay premium wages or hire personnel with more basic qualifications. An awareness of the economy and the economic drivers of any outsourced service provider is an essential component of the management oversight of these relationships.

In rapidly changing industries, outsourcing provides a buffer that protects the company from a continuous cycle of training and retraining. This is particularly significant if the requisite training is applicable only to an isolated work area or work group within the company. When the function is outsourced, it becomes the contractor's responsibility to keep the staff trained. Contractors know this and, in their fiercely competitive environment, are prepared to provide training to their employees. It is as important for them to maintain employable contract employees as it is for the firm to obtain skilled employees.

Outsourcing is also an appropriate staffing strategy for companies in highly volatile industries. Defense contractors have long used outsourced and contract employees to fulfill government contracts. Their use softens the effects on the individual company of the industry's expansions and contractions. Today, companies are adopting outsourcing as a strategy to protect themselves from large severance and other downsizing costs. Before deciding to outsource, however, a manager should consider the offsetting disadvantages.

Outsourcing Pitfalls: Beware of Losing Sight

When a company outsources a function, it frees its own management from direct oversight of an area. This transfers considerable control to an outsider. The company in effect turns over the keys to the kingdom to an outside force. Friction will arise when there are fundamental differences between the firms relative to the goals, values, and expectations. This can reduce productivity and create a hostile business climate. This is avoidable. Just as you should understand the economic driver the outsource contractor faces, so too you must make sure that the contractor clearly understands your business philosophies and your operational needs.

Some firms have found that the cost savings from outsourcing are smaller than originally estimated. This is a result of errors—oversimplification—in initial calculations of the markup and the administrative costs associated with the contract. Some firms presume that when they outsource a function, they can simply abandon having to manage the function. Just because a function is outsourced does not mean that the firm can totally eliminate its administration and management.

When a function or area grows rapidly, there will be a greater need for managerial participation. This often occurs during times of business expansion when the firm's management staff is being pulled in multiple competing directions. When management has been abstracted from a work area for some time, it is more difficult for it to accurately assess its needs and respond. This can result in the firm's placing even more dependence on the outsource contractor and can even put the firm at risk. This abstraction can make it difficult for the company to develop clear expectations and performance measures. Clear performance measures are instrumental to outsourcing success. They become the benchmarks for measuring productivity and interpreting success.

The Leased Employee: More Control

Recognizing the need to maintain more managerial oversight, some companies have chosen to lease employees rather than outsource the department's function. In these situations, an employer will contract with an contract staffing firm to provide employees to fulfill specific functions on a long-term basis. This allows the firm to maintain more control over the department's activities while eliminating many of the human resource headaches.

Health care institutions have moved to this option to meet their staffing needs in key areas such as emergency rooms and physical therapy. Many hospitals have found the difficulty of staffing their emergency rooms with trauma specialists too daunting. Strategically, the hospital needs to keep the emergency room open and admitting patients into the hospital, but recruiting and retaining staff for the emergency room is an ongoing and often difficult task. Instead they contract with an contract staffing firm to provide emergency room coverage. The contract staffing firm is the employer of record and handles recruitment, hiring, payroll, administrative paperwork, malpractice insurance, and other regulatory matters as well as managing the doctor's benefits. The hospital keeps its emergency room open, and the patient is seldom aware that treatment is coming from a leased employee. Since treatment is provided in the hospital's facility using hospital equipment, the patient's perception is that service is provided by the hospital itself.

The same scenario transfers to other industries. In these lease arrangements, the majority of the staff are former employees of the client company. For a fee, these employees are placed on the employee leasing firm's payroll. The firm then leases the employees back to their former employer on an ongoing basis. The employees are in effect supervised and managed by the customer. For example, a firm leasing its design engineers will be responsible for managing and supervising the work of the designers leased to it. The client company retains a supervisory responsibility. This can have both a positive and a negative spin. On the positive side, management can maintain a closer watch on strategically sensitive areas without having to worry about staffing and administering the human resource side of the work. Management can concentrate on accomplishing results rather than worrying about keeping the area staffed. For this reason, this staffing option has appeal to smaller firms that find it cost-effective to lease employees rather than hire the human resources personnel to support these areas.

Another similar option is "payrolling." In these instances the company recruits the worker and then asks the staffing firm to hire the person and assign him or her to perform services with the company. This option is best used when a company has a highly specialized need for which it is uniquely able to screen applicants for the required skills. This option also is used to extend service on individuals whose required retirement dates might cut them off in the midst of a project. By payrolling the individual, he or she can continue to work until the project is completed even though it extends past a company-directed retirement date.

On the not-so-bright side, firms considering leasing or payrolling as options should be aware that in some cases, the employer and the leasing company may be considered "joint employers" and share liability for the actions of leased employees. This may not seem like a significant issue until you consider the potential ramifications of a leased employee involved in a harassment or discrimination situation. Even more chilling are the implications of a leased employee involved in the ugly reality of workplace violence—whether as perpetrator or victim. How and when both agency and employer are "joint employers" is the stuff of state-by-state case law. It is an aspect, however, that all managers should be aware of. Leasing employees may reduce some of the headaches of human resources administration, but it does not eliminate the need or the responsibility for managing employees.

The advantages of leasing employees are that it provides a long-term answer for meeting specific staffing needs and transfers many costly human resource tasks and costs to the staffing provider. The contracting firm knows what level of support they can expect and the cost. The firm then has a foundation for planning that presumes a functioning department. With some very hard to staff areas, the ability to even keep the competency available can be strategically limiting. A lease arrangement reduces this constraint.

The contract agency is also a beneficiary in that it can often provide its quality employees longer-than-typical assignments. It provides greater control of their career movement. Both the contract staffing firm and its employees know the terms of the lease, which are often multiyear, and the scope. The lease also provides a clearer picture than a typical contract relationship where the contract staffing firm may have knowledge about only a small portion of the contracting firm's total human resource needs even in a single area. For the staffing firm, there is the knowledge that if the lease is renewed regularly, there is a steady revenue stream and the ability to provide a work situation with near-permanent stability for employees in a supposedly impermanent career pattern.

Successful employee leasing requires honesty between both the employer and the staffing provider. Both sides must not only communicate during the lease negotiations but also throughout the terms of the lease. Any plans for growth or strategic redefinition must be shared quickly and in confidence. This will give time for an adequate response. The contract itself should provide enough flexibility to allow both organizations to respond to changing needs. Employee performance and expectations must be communicated freely and often. It is unwise to consider leasing as a staffing solution in an atmosphere of distrust. Selecting the right provider is of the utmost importance.

For the leased employee, there are also advantages and disadvantages. Many contract employees find that leasing adds stability to their somewhat itinerant work life when they can count on staying with a leased department for several years. Some leased employees have found that their tenure with the firm leasing them can provide near-long-term job stability. Stability is of course dependent on the terms of the lease and the quality of the individual's performance. For some this tenure potential coupled with the possibility of changing employment situations without changing employers is very attractive. The work life at one installation may not meet the employee's satisfaction. The employee can seek a transfer without losing seniority gained as an employee of the staffing firm. For many individuals this continuity of employment is valuable.

For the employee who continues to yearn for a full-time, permanent position, a leased employee status has many of the inherent drawbacks of being a contract employee without the potential of moving into a permanent hire. Since the entire department is leased, this incentive is eliminated. For the employee whose entire department is transferred to a leasing contractor, this change may be unwelcome and will often be met with distrust. The employee feels as though the company is abandoning him or her.

When making a transfer of this type, it is of utmost importance to communicate to the employees the reasons for the change and to ensure that they are treated fairly. This is particularly important if they are going to be expected to return to their previous employment site without the status of being with the employer. They will frequently view themselves as second-class citizens ousted from their jobs by a greedy employer. They will act and feel as though their careers have been sacrificed. This is particularly true for employees who have considerable seniority with the firm. Their loyalties do not readily transfer to the staffing firm, and they sometimes harbor resentment against their former employer for foisting the change on them. It presents a difficult interpersonal challenge.

The Independent Contractor: A Legal Morass

Of all of the staffing options discussed in this volume, the independent contractor is most poorly understood and often misused. Whereas the contract employee is not an employee of the contracting firm, the contract employee is *in fact* employed by the contract staffing firm. As the employer, the contract staffing firm is subject to all of the state and federal regulations affecting full-time employees. These include wage and hour laws, employment of minors and nationals, the Americans with Disabilities Act (ADA), all civil rights acts, Federal Labor Standards Act (FLSA), Employee Retirement Income Securities Act of 1974 (ERISA), National Labor Relations Act, and other directives affecting state and/or federal contractors. The contract staffing firm pays the individual's social security taxes, and the workers' compensation and withholds appropriate state and federal income taxes.

The independent contractor, just like the contract employee, is not an employee of the contracting firm, and the employer pays no social security, workers' compensation, or unemployment insurance. Similarly, the employer does not have to meet other regulations and pay taxes. The independent contractor is in fact assumed to be working for himself or herself. This generates the confusion. The IRS estimates that 38 percent of those paying taxes as independent contractors are in fact employees. The large number of misclassified employees is of concern to the IRS and the Department of Labor. To these government entities, independent contractors represent significant lost potential revenues and numbers of unprotected workers.

Many employers have tried to enjoy the financial advantages of using this type of employee without recognizing that the true independent contractor must meet a stringent 20-point test established by the IRS. Some elements of the test are indeed subject to interpretation, but in 1995 the IRS stepped up its enforcement and reclassification of those not passing the test. Continued federal interest can be anticipated as the government responds to the changing reality of the American workforce.

The pitfall for employers occurs when the IRS reclassifies an employee. The IRS holds the *employer* liable for outstanding payments and substantial fines. The IRS does not care how the employer and the independent contractor have structured their agreement; the IRS simply asks if the employee met the 20-point test. The IRS considers any yes answer to the questions given below to be evidence of an employer-employee relationship. Here are the 20 points the employer must be able to answer no.

1. Do you provide the worker with instructions as to when, where and how work is performed?

2. Did you train the worker in order to have the job performed correctly?

3. Are the worker's services a vital part of your company's operation?

4. Is the person prevented from delegating the work to others?

5. Is the worker prohibited from hiring, supervising, and paying assistants?

6. Does the worker perform services for you on a regular and continuous basis?

7. Do you set the hours of service for the worker?

8. Does the person work full-time for your company?

9. Does the worker perform duties on your company's premises?

10. Do you control the order and sequence of the work performed?

11. Do you require the worker to submit oral or written reports?

12. Do you pay the worker by the hour, week, or month?

13. Do you pay the worker's business and travel expenses?

14. Do you furnish tools or equipment for the worker?

15. Does the worker lack a "significant investment" in tools, equipment, and facilities?

16. Is the worker insulated from suffering a loss as a result of the activities performed for your company?

17. Does the worker perform services solely for your firm?

18. Does the worker not make services available to the general public?

19. Do you have the right to discharge a worker at will?

20. Can the worker end the relationship without incurring any liability?

The IRS represents only one interested federal party. The Department of Labor also has a vested interest in the independent contract; however, FLSA uses a similar, not identical, list of factors in classifying employees as independent contractors:

1. The degree of control exercised by the employee

2. The extent of the relative investments of the worker and the employer

3. The degree to which the worker's opportunity for profits and loss is determined by the employer

4. The skill and initiative required in performing the job

5. The permanency of the relationship

When the Department of Labor, using its factors, determines that there has been a misclassification, the employer immediately becomes liable for overtime pay that the employee was entitled to. There are two strategies for reducing the potential bite of regulatory enforcement: documentation and hiring through staffing firms.

Any understanding of the relationship between the independent contractor and an employer must be covered in ample documentation. The services of a labor attorney are recommended in developing an agreement with an independent contractor. The attorney can prevent costly mistakes from occurring. The second strategy that an employer can use to safeguard against the pitfalls of misclassification is to hire the work done through a contract staffing firm. The employee works for the contract staffing firm, and the firm bears the responsibility for meeting the regulatory demands. The reduction of risk and not having to file the IRS forms (income is reported on a Form 1099 instead of a W-2) may in fact render the firm's fees a bargain.

There are situations when an independent contractor is an appropriate staffing solution. Some companies have found that it is a solution for obtaining the services of recently retired employees. The individual must be able to work without specific supervision and meet the 20-point test. As the IRS continues to place these workers under its regulatory microscope, firms find it increasingly advantageous to hire these staff back through a staffing firm. In this instance the employee does not have to sustain the cost or endure the paperwork of being a small-business operator and can instead perform the requisite tasks.

The decision whether to use contract, outsourced, leased staff, or independent contractors to meet contingent staffing needs will vary with the type of work, the availability of existing permanent staff, and the type of project your company handles. No matter which solution a manager chooses, the expected benefits will accrue only if the right solution is chosen and executed well. The next chapter will look at some of the managerial considerations of executing whichever strategy is chosen.

Summary of Key Points in This Chapter

1. In developing a cost analysis of contract versus permanent staff, factor in wages, benefits, productivity, and training costs; then evaluate other strategic considerations such as labor availability and the finiteness of the project.

2. Advantages of using contract employees are flexibility and cost savings; however, reaping these benefits requires managerial skill.

3. Project selection is key to achieving cost savings. Key to reducing training costs are controlling the need for contract employees to boundary span and their need for company-specific information.

4. Managers must develop accurate skill inventories to hire contract staff effectively.

5. Outsourcing may result in loss of control and perspective. Outsourcing does not eliminate the need for managerial oversight.

6. Leased employees provide more control while reducing human resource–related administrative costs.

7. Using independent contractors presents special legal challenges and should be undertaken carefully.

6

Teams, TQM, Work Groups: Integrating the Contingent Hire

It is 4:30 p.m. on a Friday afternoon. Jack, the assistant manager of the design department, is just getting ready to leave to join the rest of his department after a lengthy, but essential, phone call. The department knocked off work at 4:00 p.m., an hour early, to celebrate the conclusion of a major piece of their work. As Jack leaves his office, he sees a lone PC glowing. He realizes that it is where Mike, the contract designer, works. Jack peers around the corner and sees Mike working diligently at his computer. To Jack, he seems just a little forlorn and slightly dejected. As he approaches, Jack asks, "What are you still doing here? Why aren't you out with the rest of the guys?" He has trouble hiding his surprise when Mike replies, "I wasn't invited. You know, I'm not really a member of the department. I'm just a contract employee. I've just got a little more to do, and then I'll be heading on home."

Jack, unsure of how to handle the situation, decides not to intervene and bring Mike with him. He leaves perplexed at how this oversight could have occurred. Mike worked shoulder to shoulder with the rest of the team throughout the project. Why wasn't he included? Was it an oversight, an administrative mishap? What kind of message does this send Mike and the other team members? Was there something he didn't know or should have done?

Managers have always had the challenge of how to lead, motivate, and control the staff they supervise. In the past many managers could rely for leadership on the force of their own personalities. The loyalty they could personally garner from their employees was an important motivating factor. Basic management texts have typically included lengthy sections on the value of referent power—that power given to an individual by others. Managers were urged to develop their own managerial skills to increase their ability to relate to their employees in a way that increases their referent power. This relationship was usually built on a sense of kinship with their work units.

Managers were supported by a network of other managers committed to meeting their unit and organizational goals. The people they supervised and worked with were usually very similar to them in background, values, and personal goals. Since the employees often felt their own success was intertwined with the organization's success, a set of shared values driving toward similar goals glued the entire unit together. Employees knew that if the organization prospered, so too would they. Since careers were often shaped within the context of a single firm, its long-term health was of interest to all the employees. With the changes in both the societal outlook toward corporations and the changing workforce, these motivating factors are declining in effect. This is creating a new, more difficult challenge for managers.

A Changed Context

For several years now, the press has decried the death of the traditional employer-employee contract. With the massive downsizings and recession-generated layoffs, even those corporations like IBM that were most stalwart in guarding the contract have joined the downsizing ranks. This has given even more credence to the cries of the death of the implied social contract between employers and employees. For the technical supervisor, these downsizings have left a legacy of distrust and in some instances hostility. Many cynical employees know that the manager's job is as vulnerable as their own. They are now increasingly reluctant to tie their success to their relationship with one specific individual or even a single company.

Further, managers must juggle a workforce that includes a broader diversity of workers than at any time in the past. Within the same engineering group, the supervisor may encounter workers of differing gender, backgrounds, expectations, and personalities. The contract or contingent worker is just part of a changed mix. Today, an engineering group may include not only permanent and contract workers but also women, minorities, and the disabled. Each of these groups present the manager with a range of challenges.

The face of management itself continues to morph and reflect the diversity of the workforce. For the senior manager, this presents a very different situation than experienced by the manager of just a generation ago where the other managers reflected themselves. Adding to the stress on managers today is a changing approach to work. Today, the focus is on teamwork and employee self-direction. The trend today is to delegate more decision making and control of projects to those employees actually doing the work—empowering them to accomplish the organization's goals. This has altered the face of supervision and how managers must reward and evaluate employees. When the workers are contract as opposed to permanent, this further alters the employment paradigm.

Supervisors and employees are also focusing on processes and outcomes. In the past the supervisor was in a command and control role. Today these individuals must be secure enough to confidently guide the processes and secure the outcomes. The supervisor is now often the architect of the systems and must monitor how his or her subordinates use them to achieve the outcomes.

Frequently, the supervisor is working with declining personnel resources as every effort is made to increase productivity and reduce cost. In the changed organization of today, there are fewer human resource staff members to provide backup for the manager in handling personnel issues. Strong staffing skills are essential for the manager in these situations.

The rapid changes in technology are constantly forcing changes in processes and systems. The supervisor is expected to implement these new systems and ensure that the work group adapts to the new productivity-enhancing technology with the minimum of process loss at the outset. The supervisor can expect to regularly redesign and refine systems and processes to improve productivity while adapting a varied and changing workforce that must work with increasingly complex technology.

Challenging the manager's staffing and interpersonal skills is the management of teams that include workers with differing relationships both to the company and to the work itself. Contract and regular employees are expected to work together and achieve the organization's outcomes without the manager's constant supervision. With the advent of teams that are more self-directed, many companies have chosen to increase the span of control of their managers, further weakening their connection with the work and the workers. In organizations that use a team approach, it is assumed that all of the employees share a common goal and want to do their work. Commitment to the company's mission is part of the culture of the company. The challenge is how this fits with the changes in how work is conducted and the altered composition of the workforce itself.

The availability of reliable technology has dramatically changed how work is performed and by extension how it must be supervised. In even the smallest engineering firms, computers gather information and rapidly feed it to supervisors and other information users. The quantity and quality of the available information has increased and improved. Computer technology today has automated and controls work processes that previously required numbers of production workers.

Today, computers can readily transmit information that once required a hierarchy and network of managers. It is no longer necessary or even appropriate for organizations to maintain managers to simply provide a conduit for information flow. With the resulting flatter organizations, there are simply fewer supervisors to handle the personnel, processes, and outcomes.

The manager is not the only challenged entity. The permanent worker must also be flexible and highly adaptable. This employee no longer can nor does assume a permanent relationship with the employer. The worker also must adapt to relentless changes in technology that create a constant sense of being just barely ahead of the curve even for the most energetic and zealous. This same worker, once accustomed to being judged on his or her own work, now has to adapt to sharing responsibilities with an interdependent team. This new sharing comes during a time of heightened discomfort at job stability and other elements of a social contract once taken for granted.

The same worker is asked to relate on a par with contract and contingent workers. Many perceive these contract workers as a potential threat to their jobs. They may perceive the contract worker as filling a permanent worker's job. The contract workers are a constant reminder of the new workplace and an unpleasant reality.

To survive, the permanent staff has had to replace organizational loyalty with personal self-interest and career self-management. The result is a worker less trusting of the work situation, less loyal to the supervisor and the work unit. Many workers display characteristics that are very unattractive to the seasoned supervisor and make for difficult supervision.

For example, the Generation X employee presents a generational challenge. Many of these employees under 30 years of age do not share the values or aspirations of previous generations. As these employees increase as a percentage of the workforce, then managers can expect even more challenges. The challenges magnify when the employees are also contract workers.

Members of Generation X have watched their parents' aspirations and often lifestyles change as a result of America's competitive struggles. They do not understand why they are not automatically entitled to success and happiness. They are impatient with having to wait for the future to unfold. They have heard that their generation will be the first to do less well than its parents and are eager to beat the odds. Their posture toward the world of work is often one of hostility and extreme self-interest. They do not exhibit loyalty or interest in or enthusiasm for the careers of previous generations. Their disdain for their elders is often palpable. They are also the first truly digital generation. Reared on computers and the rapid transfer of information, they are impatient and expect near-instant results. This generation's self-interest and hostility provide their supervisors unique challenges.

Because of the economy's slow growth in the early 1990s and industry's demand for a more-skilled worker, there were few entry-level jobs available. As a result, many Generation Xers were forced to accept jobs that they considered below their training and expectations—the so-called McJob. In some instances these are the very employees that have taken contract positions as an attempt to get their foot in the door of a potential permanent employer. For these workers, the supervisor can become the embodiment of a system that they both dislike and distrust. Managing these workers whether as contract or permanent workers will continue to present a challenge both now and in the future.

Managing within a Difficult Context

To manage the contract worker in the new workplace reality, the supervisor needs to consider the types of people who seek contract employment and then determine what will motivate them. Technical employees fit a slightly different model than the typical contingent; however, their attitudes toward their work and working situation parallel other contingent employees. Many typical contingent workers in clerical or light manufacturing jobs are either between jobs or juggling multiple commitments such as schooling and rearing young children. There is also a large group of contingent workers who have completed their careers and are using contingent work as a means of easing off into retirement.

The contract engineer is not likely to fit this profile. Although many contract engineers come to contract work as a result of layoffs or other disconnections, there is a growing cadre of individuals for whom contracting is a true career. They are seeking a flexible employment option that lets them control their own careers. Many have accepted the risks inherent in contract employment because they have found that they can earn more working in contract work than they can as a direct hire with a single corporation. They realize that they will have to accept the risks of potential periods of unemployment, but they also recognize that today this a reality for all workers.

With the ability to obtain benefits and accrue retirement through a contract staffing agency, they are no longer temporary employees of these agencies. They are often long-term employees of their contract staffing firms with years of experience in working as contract workers. These employees are driven by many of the same motivators that prompt the permanent employee. They are the contract employee of choice. They represent, however, an elite group. As more engineers realize the earnings potential of contract work, their number are expected to grow.

For other contract technical employees, contract work is a stepping stone to work as a direct hire. For some it is a source of income during a protracted job search. Others, typically younger workers, use the contract situation as part of their personal career strategy. Contracts give them an opportunity to make contacts with oh-so-valuable potential employers. For younger and less skilled contract employees, contracting provides valuable experience and the kinds of job skills needed to secure a permanent job. It can also provide them a bird's-eye view of what it might be like to work for a specific company or at a specific type of work. These workers knowingly use their contracting experience as a trial run at a company or occupation. There is no single reason that typically motivates individuals to seek contract employment. In managing contingent workers, the supervisor needs to bear in mind all the reasons that individuals seek this type of employment and understand how this applies to the individuals in their work unit.

Although it is important in determining how to lead and motivate a contract employee to understand the types of situations that put them in this work, it is equally important to recollect why individuals pursue technical careers. Engineers in particular are problem solvers. They often find the most job satisfaction when they are deep in solving a complex technical puzzle. Their strong dependence on numerical data for decision making also means that the engineer is usually very aware of the numbers that guide decisions, particularly those influencing their careers. They are precise and prefer to quantify rather than intuit results.

Given their interests and training, technical staff are less likely to focus on the softer aspects of the work environment. The nice-looking office is less of a driver than a fascinating assignment. Similarly, they are more likely to consider the assignment itself in determining a job preference with equal tangible rewards. Contract engineers are also well aware that they are being brought in because they can provide a specific set of skills needed by the organization. They expect challenges.

Since many contract engineers have held permanent jobs at some point in their careers, it is important to remember not only the engineering mind-set but also the engineer's expectations of a job. They are engineers first and contract employees second. As more individuals pursue contracting during their career, it will be easier for managers and coworkers to empathize and understand the psychological impact of changing from permanent to contract work. In one instance, a career is built on history and long-term relationships; for the other, it is predicated on change.

The difficulties individuals have adapting to this type of career can actually present the manager with advantages. Because it is a new career modality, many contract employees have a difficult time bifurcating their loyalty between the staffing firms and their contracts. They work for a staffing firm. This is in fact where they should place their personal loyalty and commitment. Surveys, however, have shown that many actually are more loyal to the firms contracting for their services than to their staffing firm employer. They still place their loyalty where they perform their work. This makes them relatively easy employees to manage and motivate. They readily assimilate the organization's goals and will work to achieve the expected outcomes.

Enhancing the Positive, Driving Out the Negative

The easiest way to understand how to motivate a contract employee is to understand elements that might detract from their job satisfaction and lead to poor, subpar, performance. Management theorists have shown that disincentives can be as strong as those positive elements that create motivation. Given the contract engineers's basic personality and that most contract employees are willing to develop strong connections with their contractors, it is important to consider the elements within the contract situation that could break these bonds. The manager can then concentrate on eliminating dissatisfying factors and enjoy a more positive result. These factors for the most part tie directly to how managers structure the work situation and treat the contract employee.

Remember the contract engineer is human. Organizational research has shown that many temporary or contract workers are discouraged by the dehumanizing and impersonal way they are treated on the job. This treatment is often a result of the manager's not recalibrating his or her own expectations of the employee. In the past, temporary help was short-term and usually was used just to fill in while permanent employees were ill or on vacation or when a backlog of work required extra help. The firm's investment was minimal. Many managers felt that the less managerial effort expended during the transaction the better. The ideal "temporary" was hired with a phone call, put to work, and not thought about. Everyone would heave a sigh of relief when the permanent person returned or the workload returned to normal levels—another crisis averted.

In firms where there is a history of this type of hiring, managers will often treat contract employees as if they were going to remain with the firm just a very brief time. They maintain the temporary model. Since the average technical contract is now extending to at least 6 months, this is an erroneous assumption that can lead to poor performance. These workers are not 2-week temporary fill-ins. The contract technical staffer will be around for a significant stay.

Contract employees also complain that they are treated as if they are invisible. This leads to a sense of inferiority, which can erode self-confidence and lead to poor decision making, decreased initiative, and overall poor performance. When the employees were fill-ins, they were intended to be of minimal consequence. They were to maintain a status quo or work until a specific backlog of work was whittled away. They often sat at a permanent staffer's desk or in temporary work areas and simply held the fort down until the permanent person returned or until the situation returned to normal. They had no identity of their own and were just surrogates for someone else.

When the manager treats the contract employees as if they do not exist or will somehow magically vanish in just a few days, other employees will take their cues from them, and the situation becomes degrading for the contract employee. Managers who do not recognize the value of the employee will sometimes transmit to the contract employee a sense of low worth. In many instances this is unproductive for the organization. As contract employees stay longer on their contracts, the firm's investment in them increases. The more the firm invests in the employee, the more imperative it is for the firm to maximize return on this investment. A demoralized employee does not provide good return on investment. Low morale can lead to increased need for management intervention and poor performance.

Although they are hired through a contract staffing firm, the contractor should be mindful of the cost of making the hire and the need to get return on it. The manager has spent considerable managerial time and resources in defining the contract employee's position and finding the right individual for the job. It behooves the manager then to recognize the value of this investment and treat the contract employee accordingly.

Many managers, again working from a flawed set of assumptions, will underutilize the contract employee's talents. Although they have read and reviewed the employee's résumé and even interviewed the individual prior to hiring, they will often pigeon-hole the employee into work that is too basic to provide either satisfaction or challenge. Although it is recommended that managers limit the contract employee's need to boundary span and use corporate infrastructure to decrease training curve and costs, there is no reason to retard the talented contract employee's ability to contribute. This is particularly important if the contract employee is a high-paid technical specialist. The firm may be paying a premium for the contract engineer's skills, and it is in the organization's best interest to maximize the investment.

The manager should also be alert to allowing other employees to dump unpleasant or unattractive portions of their own work assignments on the contract employee just because of their difference in status. This can actually result in the contract employee's being unable to fulfill the task the he or she was brought in to do and can increase the overall cost of using the contract engineer.

There is also a strong temptation to hold the contract employee up to a different and often harsher standard than a permanent employee. Although no one would advocate establishing a lesser standard of performance for a contract employee, many managers are intolerant of *any* mistakes or miscommunications with contract employees. In essence this puts the contract employee perpetually at a disadvantage. It is almost as if the manager is waiting for them to make a mistake or do something wrong. This creates an unnecessary level of stress for the contract worker, which can lead to distrust, poor communication, and mutual dissatisfaction.

Just as managers are often all too willing to misjudge the contract engineer's performance, they are often just as unwilling to provide those oh-so-satisfying "atta boys" that signal a job well done. Many contract employees complain that not only are they invisible until they make a mistake but they are also unappreciated. Some forms of appreciation, such as a simple words of praise for a job well done, cost virtually nothing yet pay off richly in job satisfaction and resultant motivation. The contract engineer deserves the same level of input—praise and blame—given to other employees.

Contract employees also complain about being left out and not being treated as part of the team. This is particularly detrimental in organizations that expect teamwork as the source of results. The contract employee can feel like an outsider looking in. The contract employee, like Mike at the start of this chapter, is a part of the team for getting the job done and should be included in the team's celebrations. They know they are contract workers, but they also know that they are contributing to the team's success. To exclude them is to devalue their work. This not only creates difficulties between the manager and the contract worker but it also sends an unpleasant message to the other members of the team.

Managers should exercise caution in how much they include contract employees. There are legal risks and potential pitfalls in too aggressively incorporating the contract staffer into the work group. These will be addressed in what not to do later in this chapter along with strategies for motivating the employee beyond simply reducing dissatisfying factors.

Remember, the contract employee knows that he or she serves at your behest. The contract employee knows from the start that, although working under a contract, the contract itself is with the contract staffing firm and not with them personally. The individual is constantly aware that, if you are displeased, you can simply call and notify the contract staffing firm that the individual is not working out. This may and can result in the individual's being removed from the contract and can lead to further difficulties for the individual in securing additional placements. This can add to the stress on the contract employee. As the manager, you can readily motivate them by recognizing the tenuous nature of their work and not exploiting it.

Don't offer false promises or raise false hopes. Many contract employees are ambivalent toward contract work. Many want a long-term, stable job. It is human nature to read heartfelt desires into a transaction. When you interview the contract employee before hiring, don't suggest that the position could become permanent if this is just a remote possibility. It raises unfair expectations that in turn lead to disillusionment. Similarly, you want to provide the contract staffing firm an honest appraisal of the job's potential so that your contract staffing firm contact does not transmit false information to the employee. If the job is going to be a long-term assignment, identify it as such, but do not hold out false hopes of permanency. Further, for the long-term contract employee who has chosen contract employment as a career, this blandishment is meaningless. The individual is not about to change a lucrative way of life for a long-term so-called stable job. He or she already has a stable job with a contract staffing firm.

The manager should also keep in mind the contract employee's work environment. If you provide a pleasant workplace and interesting work, the employee will actively seek tasks and assignments that may extend his or her time with your organization. Managers should be aware of this and address the issues it presents. If you need the individual for longer periods, you should rapidly transmit this to the staffing firm, or you may find the individual's contract ending and the firm ready to reassign him or her before you are ready to end the relationship.

Similarly, it is important to provide a realistic job preview for the employee. Every job has more and less attractive components. Managers who fail to describe the less attractive aspects of a job provide an unrealistic impression that can lead only to job dissatisfaction. When reality hits and there are numerous unpleasantries, the employee feels lied to and becomes distrusting. The contract employee is no different than a permanent new hire in this respect. Although there may have been no malice of intent on your part, the result will be the same.

Keep in mind the individual's experience and capabilities. Many contract employees feel underemployed. This is particularly true of the most seasoned worker, displaced from permanent employment. These employees have often had jobs that presented them challenges and even a promise of a future. Engineers in particular are problem solvers. Too easy a challenge will leave them wondering if that it all that there is to the job. Many are highly productive workers and eagerly await whatever challenges you can present them. The manager's task is to keep them focused and on track.

A manager may well have read an individual's résumé and probed qualifications prior to bringing him or her in as a contract employee. Then, it is as if manager forgets everything they learned about the experiences and qualifications that intrigued the manager enough to want the engineer.

Give the engineer work that is commensurate with his or her skills and experience, and expect the same results as from a permanent employee. Then, let him or her work with the same level of autonomy that you would provide for a regular employee. Remember, the individual is trained, skilled, and experienced to work as an engineer or as a skilled technician. Just because he or she is working for you on a contract basis, there is no need to provide work below his or her skills or to hover nearby, micromanaging every move. This is demeaning and suboptimizes the skills and capabilities of the individual.

Having Your Policies and Procedures in Place before You Hire

Although it is all well and good to keep in mind that the contract engineer is human and expects honest and fair treatment, it is much easier to actualize this if the policies and procedures for handling contract employees are in place before they are hired. Some companies will already have specific procedures for handling the actual hiring process. There are requisition forms to fill out and procedures for handling the preemployment paperwork. These are not the only preplanning elements that the manager needs to consider or has control over. Here are some planning strategies that will make it easier in the long run to manage the contract engineer.

Develop an accurate description of the job. It is very important that you develop a clear picture of what you envision the contract employee accomplishing. Chapter 9 and the appendix provide guidance on how to develop a job order that will reflect your department's needs. It is important to focus on the results that you will want. The contract hire is an employee from whom you expect specific results. Given that contract assignments are time-bound, it is very important that you also develop a picture of how long you expect it will take the individual to achieve the results and progress measures. This will allow you to give the contract staffing firm and the prospective contract employee an accurate picture of the length of the assignment. You should be able articulate this so that the staffing firm clearly understands the tenure and potential of the position.

Be prepared in advance for questions about whether the job will lead to a permanent position. Don't just guess or give an estimate. This is a very important question for the staffing firm and its employees. Under no circumstances should you falsely dangle the prospects of full-time permanent employment as an incentive to get a "better" employee unless it is truly there. You will want to check to see if your company already has a policy that provides contract employees a preferred status in applying for permanent positions or just puts their names in along with other applicants. This will allow you to avoid setting false expectations for the staffing firm and its employees.

Similarly, if recruiting highly skilled technical staff has been a problem and you intend to attempt to convert the contract hire into a permanent employee, you should let the contract staffing firm know this up front. Concealing this intent could be challenged as unethical. The staffing firm has recruitment costs and certain expectations of the employees' tenure with it. Most staffing firms will work with you to convert a contract position to a permanent hire.

Verify if there are in place personnel policies that pertain to the treatment of contract employees. The contract employee is frequently treated in a degrading and dehumanizing manner by both managers *and* their subordinates. You may treat your employees sensitively, but you will want to know what provisions the company has made for protecting the dignity of these workers. In companies that have extensive experience with using contract employees, there may already be in place personnel procedures that directly address the rights as well as the day-to-day treatment of the contract workers. If your department is the bellwether as your firm moves into using contract engineers, you may want to develop a set of policies for the treatment of the contract employee that could be used throughout the firm.

This may seem at first unnecessary, but having a policy in place may serve to decrease your liability in the event of some misconduct, such as sexual harassment or discrimination, involving a permanent and contract employee. The task may just require extending existing policies to reflect the contract employee's specialized work situation. It will also help you formulate how you will ethically and respectfully treat these workers.

Determine in advance orientation procedures. Before the employee arrives, you will need to have determined and arranged for an orientation. The better oriented the employee is, the faster he or she will become a productive member of the department. Beyond outlining the specific task at hand and the work area, there are obvious elements of any orientation—introduction to coworkers, work hours and rules, restrooms, cafeterias, and coffee and refrigerator protocols. Believe it or not, some contract employees complain that contracting firms will even neglect to cover these basics.

Orientation should also consider some of the realities of today's work environment. Today's employees are no longer just working at a desk with a phone. Our society is both digital and highly interconnected. The employee's workstation now includes sophisticated communications devices and more often than not a computer connected to a network. Although you expect to hire someone with experience in using these technologies, your orientation should include bringing the person up to speed on your technology's custom features.

Just because the contract employee is technology literate, do not expect him or her to be able to work instantly at a terminal and with a network system that your information systems department has spent 10 years customizing. The person may be a whiz at using a software package, but you must make sure the individual understands your system configuration and how to access it. It is very easy to forget that an outsider won't know all of the special quirks of the system you have worked with every day as it evolved. For the newcomer a shared printer hidden away in a remote area of the department can turn a simple one-page memo into a hide-and-seek adventure.

Telephone systems are rapidly becoming as sophisticated as computer systems with voice mail and conferencing features almost standard. Regrettably, these systems too are highly individualistic. Although "1" is usually yes and "2" is no, the person should not have to listen to all of the prompts to conduct business. Fax machines and e-mail systems should be included in the communications orientation.

If your organization relies on e-mail for some types of communications, fax for others, and voice mail for still others, be sure to consider how you will orient the contract employee on when to use each mode. For some firms, different types of communication are relegated to each type of transmission. If it is standard procedure to follow fax transmissions with a hard copy, or some other protocol, do not expect the contract employee to somehow magically know your business's routines. Because of the newness of these technologies, much of the protocol is unwritten; it is just something everyone knows except the new contract hire.

Determine in advance the contract employee's impact on the department. Managers should not overlook the psychological impact of the contract worker on the permanent staff. When the manager has made adjustments in workload to minimize the contract engineer's need to boundary span or use extensive proprietary knowledge, he or she should also consider whether this has resulted in sloughing of harder more complex tasks onto permanent employees already working at or near capacity. If this is the case, resentment may result. Before you bring the contract employee on board, determine how you will handle necessary workload adjustments without creating potential areas of conflict.

Similarly, if the contract employee is brought in to fill a perceived permanent vacancy, the staff may resent the hire. For example, if a contract engineer is brought in after a recent retirement, the permanent staff may perceive the individual as filling the retiree's position even if the workload and the situation has changed. You will need to articulate why you are filling the position with a contract employee. If the work expectations are different, make sure the permanent employees understand them and do not just assume that the individual will be doing the previous person's job.

Also, the permanent employee may develop a negative profile toward management and the organization if he or she perceives that contract employees are either exploited or filling jobs that should be part of their workforce. The manager will need to consider the potential impact in advance and what strategies must be taken to soften any possible negative perceptions.

In organizations that rely on teams or self-directed workgroups, the integration of contract workers with permanent staff is tantamount to the success of the contract hire. Before hiring a contract employee, consider what type of work situation you use. This will help you obtain an employee who will meet your needs. The contract staffing firm needs to know what type of workplace you operate. If you are prepared in advance, the transition will be smooth. Many contract staff have extensive experience in working with the new management schematics. You can with a little forethought maximize the transfer of their experience into your organization. When you recruit from a contract staffing agency, you will want to make sure that you understand the types of work situations each candidate has experience with and choose the employee that will most benefit your organization.

Keeping Your Contract Employees on Track and Performing

Contract employees are just like any other employee. They come to work each day and do the same or similar tasks as your permanent employees. The real differences are in their professional long-term outlook and how you, as the manager, fit in the mix. Given that contract employees are usually highly motivated and dedicated to their contracting firms, they do not present a problem in motivation but rather one of maintenance of effort and direction.

The complaint heard most frequently from contract employees is that they are treated as though they do not belong; thus you will enhance performance when you ensure that the employee is treated with the same dignity and respect provided your permanent employees. If you make an effort to show that the contract engineer belongs and is making a valuable contribution to the effort, you will most probably find yourself rewarded with continued top-level performance. Here are a few suggestions for delivering this message:

1. *Never refer to the engineer as a "temporary."* Similarly, the more you dwell on his or her status as a contract employee and single the individual out, the more difficult you will make the situation. If you continuously single out and treat the contract worker as different, the more alienated the worker will become.

2. *Make sure contract hires receive enough orientation so that they can rapidly perform their function immediately and adequately.* Make sure orien-

tation occurs in a timely fashion. The hire on a 6-month contract should not receive orientation in the middle of month 3. This might be fine for an employee you expect to stay with the company for a career but not for a contract staffer.

3. *Avoid the temptation to overmanage the individual.* This goes directly back to recalling that you are dealing with a professional and, as such, you can expect to have hired through the staffing firm an individual able to work with some autonomy. Remember, the individual contract employment has already shown a commitment to working out of the traditional employment model. They are often willing and able to work independently.

4. *Recognize and praise work well done.* Everyone likes to know that he or she is doing a good job. The contract employee should not be precluded from recognition when a job is truly well done. Remember that the contract employee works on a triangular relationship with you and the staffing firm. Make sure that any special recognition for the individual is transmitted to and through the staffing firm contact. The engineer will appreciate this since the staffing firm is the career anchor. When you send your positive feedback to the individual through the staffing firm channel, you also provide a level of cushioning from legal risks that could jeopardize the employer-of-record status.

5. *Provide access to all necessary information to accomplish the assigned job.* Some managers have tried excluding him or her from meetings and information loops and have thus crippled the contract employee's ability to complete tasks and respond to organizational needs in a timely fashion. In today's information-driven workplace, this is foolhardy. It will not only cripple the person's ability to function at peak but it will also negatively single him or her out. You will also find that other employees will provide the information through back channels that may distort it. Companies that are secretive usually have accurate and sometimes harmful informal communication channels, the much vaunted grapevine. Be as inclusive as you can. Follow the same basic guidelines you would for any other employee.

6. *Train as needed.* Don't just expect the contract staffer to pick up something you are training your own staff on. For example, if your organization shifts telephone systems or changes a major computer package it is using, you should include the contract employees in the implementation planning. You cannot just assume that they will pick it up from the other employees. If, for example, to make a smooth shift, the company decides to provide all staff a half day's training, you should include the contract staff. Exclude the contract employee only if

their contract is ending before the person would need to use the system. The contract engineer needs to know how to use the new system as much as any other employee. If you are planning this type of training, you will want to discuss it with the representative of the contract staffing firm. You may find that the firm itself has a policy on how it financially handles training provided for its employees by their contractors.

7. *Challenge the employee.* The highly motivated self-starter is a blessing in any organization. It does not matter if the individual is a contract or a permanent employee. The organization will benefit more from harnessing the creativity and energy of these individuals than from holding them back. High-functioning, motivated individuals thrive on challenges and self-motivate from interest in the work. Seek their input, use their knowledge, and offer these individuals an opportunity to grow intellectually. Even though you cannot provide career growth in the traditional sense, everything that a contract engineer learns while working adds to the individual's personal capital. You may also find that the contract employee, with a perspective framed on a variety of experiences beyond your company, will provide valuable insights and fresh opinions and solutions. Use this wealth of knowledge.

8. *Trust the contract employee as you would any other.* Yes, the employee has a different relationship with your company, but you will keep his or her motivation and loyalty if you demonstrate a level of trust appropriate for the job he or she is assigned to perform. Entrust your contract employees with keys to their work area, passwords needed to get information they need to use, or other symbols of trust.

9. *Include your contract engineers in social functions.* If they work shoulder to shoulder with your regular employees, they should be part of the same celebrations. Mike, the contract engineer at the opening of this chapter, was singled out and cut a forlorn figure that will leave an indelible impression on Jack. His other colleagues will find it difficult at work to recap the fun they had at the celebration knowing that Mike was not there. The situation will be awkward for everyone.

In many instances, exclusion will rankle regular employees who have accepted the contract employee as part of their team. They will see it as a signal of how management dehumanizes its workers and by extension how the company will treat them. It is just a short jump from the treatment of the contract employee to what they might expect for themselves. In an environment of distrust of employers such as exists today, it is a questionable practice to feed this engine of doubt.

The Legal Obligations of
Too Much Inclusion

If your organization regularly uses long-term contract employees, you need to be aware of the potential legal problems that can result from integrating the long-term, contract employee too deeply into your organization. The IRS and ERISA are continuing to look for misclassified employees. Their interest is in "who is the employer of record." You can violate the staffing firm's status if you bypass it in your efforts to integrate their employees into your workforce.

For example, it is not recommended that you print company business cards with the contract hire's name on them. This creates the impression that he or she is your employee and supports such claims. Similarly, any rewards and recognition should be extended through the staffing firm to protect the status. You should also extend invitations to company picnics and other parties through the staffing firm. You should under no circumstances compromise the staffing firm's position as the employer of record; otherwise, you could find your firm subject to fines and with a monetary obligation (up to and including potential rights to pension) to a long-term contract employee.

The contract staffing firm is aware of the need to maintain a clear chronicle of the employer of record. Many of the firm's practices are designed to protect this status. It will serve you well to work closely with your staffing firm contact. This is one of the many reasons it is important to choose your staffing firm carefully.

Teams, TQM, Work Groups:
Yet Another Set of Challenges

Today, many companies rely on teams and work groups to accomplish results that previously would have been individual performances. Teams are already reality in many manufacturing installations, and their use continues to grow apace with their documented success. The use of teams is often coupled with other corporate-wide initiatives in self-direction, total quality management, and ISO 900X. The same economic drivers that push these initiatives are also behind the move to a more flexible workforce.

Managers can expect to manage a workplace that is team-driven and also heavily populated with contingent workers. These may be seasonal employees, college coop students, or contract and contingent employees. Integrating these alternative employees into a team-driven workplace requires understanding team dynamics and recognizing how they are played out in your organization.

Not all teams are alike either in function or in how they perform. The type of team and the organization's level of experience in using teams may strongly influence the ease with which a manager can integrate a contract technical hire into the team structure. Natural or departmental teams require a different strategy than process improvement teams or cross-functional teams.

As a matter of definition, a *natural team or departmental team* is a group of people working within a process to assist each other on a daily basis to accomplish their goals. The *process improvement team* is typically empowered to review a "process," identify areas for improvement, and then suggest and implement the improvement. The *cross-functional team* is a group of people who represent a number related processes and have been empowered to review the system, identify areas of improvement, then suggest or implement solutions. Contract employees can work and do work within all three types of teams.

Where the team structure is organization-wide, the contract hire will by definition be part of a departmental or natural team. Their regular work group will constitute the team. Natural work teams usually have in place mechanics for integrating new hires. These are often the rituals of welcoming—introducing the individual, outlining for the team new or changed responsibilities, and orienting the new hire into how the team functions. The same rituals should be followed for welcoming new contract employees. This sends the subtle message that they are now part of the team. Whether the team accepts the new hire is often a function of team dynamics. Since these dynamics vary from team to team within an organization, the manager must be cautious when adding a contract hire to consider the team and its flexibility and resiliency.

Process teams are often the easiest to bring contract employees into. Because these teams are focused on specific problem tasks, the contract technical employee can provide a specific specialized expertise that the team needs. They are valued content experts. The team will eagerly seek their input and readily embrace them as able to help the team accomplish its mission. The contract employee is well suited for these situations. The nature of their employment is an additional incentive to using them on a team that is constituted with a task to perform. The team itself will usually disband at the end of the project. The contract employee's time-bound relationship with the company fits directly into this type of work. If success is to be achieved with minimal process loss, it is important to bring the contract engineer into these teams at the appropriate time in the team's development.

Cross-functional teams that bring together individuals with varying expertise representing a number of related processes present the greatest challenge for using contract employees. These teams will frequently need the technical expertise that the technical contract employee can bring; however, because of the contract employee's relationship with the company, he or she is inappropriate as a representative of work process areas. In assigning a contract employee to a cross-functional team, the manager should make sure that there is also a process owner on the team in as much as the contract employee can bring to the team only certain skills, not functional or process ownership. It is easy to forget this when using long-term contract employees who are fully integrated into the work situation.

Whether the team is natural, process improvement, or cross-functional, the contract employee's successful integration will in part be dependent on the teams themselves. Teams are dynamic and organic. They each have different levels of resiliency and adaptability. They also go through distinct phases of development. Managers who understand team formation and dynamics will be in the best position to judge their team's ability to accept and achieve high-level results with contract employees.

When a team is first *forming,* there is a feeling-out period as each member tries to establish a place in the team and the team grapples with identifying its task and the rules and procedures that will govern the group. Teams pass through this phase at varying speeds depending on the task at hand and the personalities and experiences of the team members. Employees familiar with team activities and team process will quickly accomplish this first phase and move forward.

When a contract employee is expected to work with a newly forming team, it is helpful for the individual to begin the process with the other team members. If you expect to use contract employees in teams, you should seek this skill as part of the specific qualifications for employment so that the contract employee can quickly transit with the team through this forming process.

In situations in which the organization is just beginning to use team structures, a contract employee skilled in teamwork and in team dynamics can actually help move new teams forward. Learning the interpersonal dynamics of teamwork takes time. This is why organizations must be patient when first introducing teaming into the workplace. It takes time to learn constructive ways of resolving conflict and how to respect and use everyone's opinion in the group. For many individuals, group decision making seems an alien and complex way to get things done. With its heavy emphasis on process, individuals who understand how to facilitate process changes can help the team stay on track and move past roadblocks. When you can include personnel with more skill in teamwork, you can shorten the learning curve as team members piggyback their skills from the more skilled members. Whether the individuals are contract or permanent staff is less of an issue than what they can add to the team's ability to move forward rapidly. Contract employees familiar with teaming and team dynamics can help pull the team along more rapidly.

Teams storm during their second developmental phase. This *storming* is a time of challenge for the manager. This is when conflicts occur as roles, procedures, and even the validity of the task at hand are hashed out. Again, the roles and personalities can either speed the process or retard forward movement. The manager using contract employees in teams must be aware that it is during this phase that challenges to the employee's role and position are most likely to occur. The more clearly the manager defines and articulates the contract engineer's role at the outset, the less likely it is that conflicts will occur. When the manager is able to bring the contract employee into the team at its outset, some of this can be avoided.

Teams will storm for varying lengths of time. While this is occurring, managers should not add new members to the team in as much as new members, of any employment status, can either intensify the storm or prolong it. Similarly, because of the conflicts that occur during the storming process, the manager needs to be patient with emerging role conflicts that impact the contract employee and not seek to remove the person from the team or intervene. These challenges are an unpleasant and transitory part of the process.

During the third phase, *norming* occurs. At this time team members concentrate on the task and develop group cohesion and the necessary interdependence to accomplish the team's goals. Members develop trust and sensitivity to one another. Contract employees, well oriented and integrated into the regular life of the firm, will become accepted members of the team. It is important as a manager to recognize the strong bonding that occurs and to be sensitive to process issues that could break the team apart. Managers should be cautious in adding new team members during this phase.

The addition, whether permanent or contract employee, can send the team back into a storming phase as the team realigns to accommodate the new person. Managers should particularly avoid placing new contract employee into a team that is just norming. Similarly, removing or reassigning team members (having a contract engineer's contract end) at this juncture can destabilize the team and return it to a storming phase as there is a shift in what the team considered established roles.

Performing occurs during the fourth phase. Groups develop strong identities, and the members actively work toward the group's goals. At this phase, the group members feel comfortable enough with each other to seek assistance freely from members of the group and those outside the group who are needed to accomplish the group's goals. Members in the group usually develop pride in their accomplishments and develop an interest in sharing their results.

When teams are *performing*, they are focused on accomplishing their goals. Many teams at this time are comfortable enough with themselves to absorb new members. There are, however, some aspects of the team's dynamics that the manager should be aware of in adding a contract employee. During the early phases of team development, the team is building a relationship, a context within which it will work. The tenure of the team and the intensity of the process, just as in any relationship, will color the context.

In many teams with a long tenure and a rich context, the members will become so aligned with each other that they develop a form of interpersonal code among themselves. The team will also develop strong adaptations to accommodate differing roles, individual strengths, and weaknesses. Sometimes these will seem confusing to the outsider. The team itself may be adaptable and resilient enough to accept new members, but unless there is a recognition of the team's history and context, the new addition will remain an outsider for an extended period of time.

As companies develop more experience with using teams, there is growing concern about how to revitalize long-standing teams that have grown complacent. Motivation levels may lag and the team's performance may fall off as the team wears in together and the first heady rush of success wears off. The contract employee with a new perspective, experienced with teamwork yet new to the company and the team, can frequently add the flood of new energy that will shake the cobwebs off the existing team. Just as with any team, the manager seeking to revitalize a team through contract employees should be mindful of the team's dynamics in bringing the new catalyst—the contract employee—into an existing team.

Adding contract staff to existing teams requires a sensitivity to the teams' developmental phase *and* its unique interpersonal dynamics. No matter how skilled the organization is in the use of teams, they are groups of people working together, and they will reflect the richness of the personalities of the members. This is the much vaunted "team chemistry" in sports, where highly talented superstars underperform because they cannot quite work together with their team mates. The manager is not unlike a coach with a strong facilitative responsibility.

Teams react differently to the addition of new members. Some teams are *self-sealing*; others are *absorptive.* A self-sealing team is so tightly bonded to itself that newcomers of any persuasion are viewed as intruders. The team may actually try to accept new members, but its dynamics, shared history, and personalities will inhibit acceptance.

Managers should avoid trying to place a contract employee in a self-sealing team. In these teams, the contract employee, however, can fulfill roles of process documentation that require an external, impartial, noninvolved individual. For example, a self-sealing process team may develop a set of procedures for improving workplace safety. A contract employee could develop the manual that outlines the processes decided upon by the team

On the other hand, an absorptive team is more able to adapt and bring in new members. These teams may be strongly bonded and high functioning, but their interpersonal dynamics allow them to adapt to change more easily. New members can join the team without disrupting the chemistry, and members can leave the team without breaking it up. The contract employee can flex into these teams almost as easily as any other employee, and the manager can confidently assign the employee knowing that the team will handle the integration process.

An organization may have both types of teams, and the manager must understand and be able to interpret the team's ability to absorb new members before trying to add them. Contract employees are ideally suited for content expertise roles in performing teams. The manager's challenge is to interpret how and when to add the individual.

Many organizations using teams have developed specific rituals for facilitating movement between teams. For some teams this may be as simple as an introduction at a team meeting. Other teams eat together or use other, often team-developed, bonding activities. If you, as the manager, expect the contract staff to function as part of the team, do not shortcut these team activities. This can inhibit acceptance and performance.

Total quality organizations have adopted a specific cultures that focuses on customer satisfaction. Adopting this culture requires an organizational commitment and adaptation. If your organization is a total quality culture, you will need to make sure that this culture is understood by your staffing firm before you begin looking for contract employees. In quality organizations all employees must share the organization's commitment to quality and be ready participants in the culture of improvement. The contract staffing firm will want to recruit contract employees that can readily adapt to your quality culture. If you seek out contract employees with experience in quality environments, you will more readily find them aligned with your organization's goals, needs, and work environment.

When adding contract staff to a quality-driven organization, the employer should develop a clear vision of how these employees will fit in the quality initiative. Management should know exactly what it expects the contract hire to accomplish. With contract engineers your organization is purchasing expertise. Part of the expertise should include experience in a quality culture. You should not have to train the individuals on the mechanics of ensuring quality. Contract staff with strong technical skills and experience in quality organizations can actually speed technology transfer within your own organization.

A quality-driven Toledo, Ohio, automotive supplier has found that contract employees are a valuable component of its staffing mix. The organization uses contract employees in its design engineering. The contract staff, working side by side with the permanent staff, was able to accelerate the company's shift from paper designs to CAD. The company experienced a rapid transfer of skills as the more experienced CAD designers, hired on contract, worked with the permanent staff.

Organizations that are seeking, have sought, and/or have obtained ISO 900X certification should follow the same advice. Seek contract employees familiar with the procedures and demands of ISO 900X. You will shorten the learning curve and ensure cultural alignment and more rapid fit. Just as contract employees familiar with teamwork can energize and enhance the team process, these employees can have the same effect when the organization is moving toward an ISO 900X environment.

With the variety of experience gained through their work with other firms, contract engineers may provide valuable insights and skills that can reduce your organization's learning curve. This may require more aggressive hiring procedures, but the payoffs will offset the efforts. There is always the potential with any employee that skills learned on one job will be transferred to another. Just as you are the beneficiary of skills gained by contract engineers at some other firms, you should be mindful that the contract employee will leave your firm with a new or enhanced set of skills. There is always a ready danger that any employee will take his or her experience to another firm. You have no guarantee how long any employee will stay with your firm. It is a risk of doing business. Unlike your permanent employees, you know when this will occur with contract employees—at the end of the contract.

Self-directed work groups present yet another challenge. Many of the same recommendations that apply to teams apply to these groups. Working in a self-directed work group requires a bias for this type of organization. If your organization depends on its technical staff to work in self-directed work groups, hiring contract staff with experience in this type of management configuration should be a priority. The quality of your staffing firm will determine whether you get employees who will complement your workforce or represent unwanted virtual aliens.

Summary of Key Points in This Chapter

1. The changed managerial context has altered the manager's role, employee loyalties, and the workforce itself. This has changed how managers can lead and motivate their employees.

2. Technology has changed how managers disseminate information, control work processes, and manage the resources, including human resources, whether contract or permanent.

3. Today's workforce, both contract and permanent, is more diverse and self-interested, particularly those of Generation X. To manage effectively, supervisors need to recognize and acknowledge the workforce's changed perspective and values.

4. Contract employees reflect the workforce. Understanding why individuals have chosen to work under contract is key to motivating them.

5. Contract technical employees are usually easy to motivate. Managers should drive out disincentives. They should treat the employee with empathy and respect, offer realistic expectations, and provide work commensurate with their skills and experience.

6. To facilitate integration of contract workers within the workplace, have policies and procedures in place before hiring.

7. Plan an adequate orientation for contract staff including how to use telecommunications and computer equipment.

8. When providing rewards and special incentives to contract workers, always work with the firm to prevent legal snarls over who is the employer of record.

9. Before placing contract engineers on work teams, evaluate the team's mission, developmental phase, flexibility, adaptability, and group dynamics.

10. Quality-directed organizations or those using self-directed work groups or having ISO 900X certification should seek contract employees with experience in these settings.

7

Selecting a Staffing Provider

Fred, the engineering manager of a northeastern plastics firm, is excited because he has just been notified that his department will be responsible over the next year for the development of two new work cells. Each will include new equipment and incorporate new materials and processes. Four years ago, to reduce costs, Fred's company downsized the engineering department. This has left Fred's department too thinly staffed to develop the new work cells without additional staff. Since both work cells are scheduled for completion within the year, Fred's management has authorized him to hire additional staff; however, only if he uses contract engineers hired through a staffing firm. The firm has never used contract engineering staff. In the past manufacturing growth was much slower and was absorbed into the regular workload of the department. The company's only experience with contingent staffing has been with a local staffing firm that has supplied temporary office help and a few manufacturing workers. The human resources manager has suggested that Fred contact this firm since it is full service and advertises its ability to provide technical staff.

Fred's first contact with the firm was friendly enough. The account executive seemed eager to meet his needs, although he seemed to have little understanding of the technical qualifications that he outlined as job requirements. He keenly needs two very skilled engineers with specific experience in plastics, work cell implementation, and fast-track projects. From his

experience in hiring permanent employees, Fred knows that
the skills he is seeking are not readily available in his area.
To Fred the account executive seemed a little too breezy and
confident about the availability of engineers and the firm's
ability to fill his job order. After hanging up the phone,
Fred's discomfort increased. He is concerned now about
whether he has made the right decision. Is this the right
contract staffing firm? Will he get the employees he needs
quickly? Should he look elsewhere?

The contract staffing industry has grown and changed in recent years as the firms watch the trends in the American workplace and react to them. Firm management is well aware of the dramatic growth in the use of contingent and contract employees and is trying to meet the changing needs of clients. Either spurred by client demand or in response to market influences, many traditional temporary staff agencies have broadened their scope. They now serve wider geographic areas and provide a richer mix of skills than previously. At the same time, there has grown and flourished a number of firms servicing specialized industries and providing specific types of staff. For example, a number of firms provide nurses and other health professionals to specifically meet the needs of hospitals that now use pools of health professionals to meet their varying patient loads. Some firms provide just accounting and finance professionals. Similarly, there has been significant growth in the number of providers that offer just technical staff with specific competency in data or information processing, design, or CAD. As with any service business, the ability and perception of how well a contract staffing firm can meet its customers' needs will vary. Only the user organization can truly judge whether a firm is meeting its specific staffing needs. This chapter, however, will present some factors that a manager, such as Fred, should consider in evaluating a staffing firm's ability to meet the organization's technical staffing needs.

When an organization decides to add staff through a staffing firm, the need is usually immediate as in Fred's case. The hiring manager more often than not has a project with a completion deadline already established and inadequate current staff to complete the work. In today's environment, the work is already severely backlogged before the manager can obtain a go-ahead to increase staff. This makes it imperative to get the right staff on site as soon as possible. Any delay in hiring the needed staff can create subsequent delays and pressures on the project and the existing staff. The manager's time is often seriously constrained in as much as ramping up a deadlined project places extraordinary demands on management. Hiring contract staff quickly and with a minimum of managerial effort is, therefore, important.

There are several factors that can influence your speed and ability to effectively hire contract staff: the staffing firm, the job description or job order submitted to the agency, and the labor market. You have control over two of these three variables. You can select the staffing firm that will best meet your needs and develop an accurate job description or job order that will allow you to secure the right employee. You have very little control over the labor market. You cannot readily alter your location or magically clone additional highly skilled individuals on a moment's notice. You have virtually no control over levels of employment. You also have little control over the cost of certain skills. When demand is heavy, you can expect to pay a market-driven price for the talent that you hire. Since you do not control the labor market, you should seek ways that will mitigate its influence on your hiring.

The labor market will also impact on a contract staffing firm's ability to fill your job order quickly. When there is significant unemployment resulting from a recession or industry contraction, the staffing firms are flooded with laid-off professionals looking for ways to meet their immediate financial needs. With a large pool of available talent, the firm can more easily recruit and select an individual with the needed qualifications. When the economy is rising or there is extraordinary demand for individuals with specific skills—such as the phenomenal growth of the Internet creating a significant and immediate demand for individuals with experience with UNIX and telecommunications—the firms experience a tightening of their sources. Their pool of available candidates shrinks, which in turn creates pressure on the price of available talent. For contract employees working in a tight job market, earnings are constrained only by the willingness to pay of those organizations needing their skills. Your contract staffing firm will be able to give you guidance on the availability and the market-driven price of the labor that you are seeking.

As a user of a staffing firm's employees, you need to be mindful of the market reality. You can soften the impact by working with an agency that provides the specific kind of talent you need on a regular basis. The specialized staffing firm will maintain a larger, more specialized pool of talent than a general staffing contractor. When you are seeking highly skilled technical staff during a tight labor market, you will need to work closely with the staffing firm to ensure that you can obtain the staff you need. You should expect these professionals to command a premium price.

The Value of Planning

Although your organization may not have used contract technical staff in the past, in all probability this will change at some time in the not-too-distant future. The worst-case scenario is to be caught with little understanding of how to use alternative staff to advantage and no plan of action for hiring. By reading this book, you are taking a significant planning step that will prevent this worst-case scenario.

Don't wait until you need contract staff to begin evaluating firms you might use. When you make the decision to use contract staff, you will want to be able to contact the staffing firm and set the hiring process in motion as rapidly as possible. Usually you need an engineer to begin working as soon as you make the decision—so why not be ready? You do not want to be just starting to look at who can provide the staff you need when you actually need the staff at work. Begin developing a list of firms that could provide you with staff well in advance of this need. When we must make a decision in a hurry, we frequently will not make an optimal choice. Staffing decisions are no different. Give yourself a chance to make a more informed decision by starting early.

Fred began his research by consulting with his human resource manager. How much help you will get from this department will depend on your organization. In some smaller companies that do very little recruiting and hiring, the experience of the human resource department with specialized contract hiring is very limited. In these instances the engineering manager must be more independent and take more initiative. On the other end of the spectrum, some organizations with large technical staffs have already developed relationships with firms to meet long-term technical staffing needs. If your organization has already built a working relationship with a staffing firm, your task will be simplified. You will need to develop a job order and work with your provider to secure the staff you need. Objectively, consider how your company approaches recruitment and hiring. Consider how many technical staff the company has added in recent past and the procedure used. This will give you an indication of how much autonomy you will have in selecting a staffing provider and filling your department's needs.

Use your professional network to gain information on staffing firms in the area. If your firm does not already have a relationship with a specific technical staffing provider, you will want to develop a list of prospective firms that provide the skills you need in your geographic area. Not only should you ask your human resource department but you should also check with other engineering managers about their experiences with staffing providers. This is one of those times when a network of professional peers is so helpful. Here are some questions that you may want to ask:

- Which contract staffing firm does the manager use?
- What type of technical staff did the agency provide?
- What skills and experience level did the manager need?
- How quickly did the staffing firm respond with candidates?
- Were the candidates qualified and appropriately screened?
- Was the manager satisfied with the employees hired?
- Would the manager use the staffing firm again?

You will want to evaluate the responses to each of these questions within the framework of your own department and needs. If the person you are talking with has only used drafters with limited CAD experience and you know that you will be needing designers with extensive CAE (computer-aided engineering) systems experience, probe more deeply about the level of skill the staffing firm is able to provide.

If your peer found that the individuals sent exceeded his or her skill needs, you may find it easier to get the skills you need. On the other hand, if the staffing firm oversells the qualifications of its contract staff, you will need to know this in advance. You will definitely want to get a balanced view of any provider. By asking your peers about their experiences, you can get a user's point of view. Any staffing firm that you talk with will want to make a good impression to get your business. It is, however, up to you to gather the information that you will need to make your own evaluation and decision.

As you do your preliminary research, make a list of the firms that you would like to personally evaluate. The range of services and the qualifications of the employees provided by contract staffing firms is quite broad. You will want to determine if the firm can provide contract employees that will meet your needs *and* whether or not you can work with the firm. If you anticipate that your workload will create a need for contract staffing in the near future, you should begin looking not only at the firms but also at the account executives you will be working with. The account executive will become your link to the firm, and it is important that you forge a comfortable working relationship.

The Account Executive: A Combination of Professionalism and Chemistry

You will want to develop a rapport and level of mutual understanding and trust between yourself and your account executive. Although the account executive who spoke with Fred was polite, confident, and professional, he came away with a sense of discomfort about the firm. To prevent feeling like Fred, you will want to ascertain the level of understanding that the account executive has of your industry and its technical needs.

Many technical staffing firms provide large numbers of staff to specific industries. Their account representatives have in-depth understanding of the special requirements of the industry. For example, ITS Technologies routinely provides staff to automotive suppliers. We are familiar with the demands of automotive design and manufacturing. When a prospective client indicates a need for a designer familiar with specific software or with the demands of a specific design component, the account executive knows from past experience with the industry how skilled the designer must be and can identify individuals who will fill the client's needs.

You will want to ask what industries the firm serves. Most firms will be glad to provide you with information about the firm and the types of clients it serves. The more experience the firm has with your industry, the more easily the account executive will understand the specific nuances of the job and the skills that will be needed to fill it. Upon request most firms will furnish references from satisfied customers. You will want to contact them.

When you talk with the account representative and the agency's references, don't be afraid to ask what level or types of staff the staffing firm provides each customer. This will help you develop a more rounded picture of the staffing firm and what you can expect. When you are checking references, be mindful of the fact that references usually reflect a special relationship with the supplier. One does not put disgruntled customers on the reference list. Assume that the list you receive has been sanitized to reflect well on the staffing firm. It is wise to probe whether there have ever been any difficulties and to ask how they were resolved. You not only want a firm that is well thought of by its references but also one that may have had to earn its stripes by resolving difficult situations.

You, however, are the only one who knows exactly what skills you require in your department, and you must be prepared to communicate them to the account executive. In Chapter 9 (pgs. 179-188), we will give you guidelines for developing job orders for contract hiring. You should not have to explain to the account executive the basics of your industry or type of manufacturing, but be prepared to answer questions specific to your organization and the positions you are seeking to fill. The more the account executive knows about your firm, the easier it will be for him or her to recruit and select staff who will fit in your organization to fill your assignments.

You will need to feel confident with the account executive so that you can communicate freely. The chemistry between you and your account executive can affect your ability to obtain the type of staff you need. If you and the account executive can build a rapport with each other, not only will you find it easier to get the staff you need but also to manage them during their assignment. If you think of the contract engineer, the account executive, and yourself as each a leg on a three-legged stool, you will quickly understand why building a strong relationship is so important to a successful contract assignment.

You should also carefully probe whether the staffing firm routinely serves businesses in your area. You may find it hard to work with a staffing firm located at a distance from your site. Although it is often possible with today's technology to communicate at a distance, you (and the employees the firm places with you) will want personal contact with the agency personnel. You should also ask whether the agency will be using local staff for your assignment, or whether the staff will have to travel some distance or require relocation. Although many contract employees will travel a considerable distance to a job or even temporarily relocate away from their families for extended periods of time, they will also usually take employment closer to home if this becomes an option. If you can get contract employees whose homes are close to your site, you may experience less turnover.

In areas with snowy winters and months of bad weather, long commutes can translate in lost work even for conscientious employees. An employee traveling 100 miles each way in the snow can become tired, dissatisfied, and eager for work that will eliminate the harrowing commute. Similarly, if the traffic in your area turns a short drive into a parking experience every day, you can assume that this will render your contract less satisfactory than another with a less difficult travel situation. Travel time and an individual's willingness to endure transportation hardships are often dictated by geographic circumstances. For example, an individual living in a small town with two stop lights will often refer to the "rush hour" as a time when traffic is waiting at both lights. On the other hand, a veteran of a major metropolitan area where traffic jams are miles long and waits are lengthy would wonder if the small-town individual was being humorous.

Since you will need to communicate with the contract staffing firm on a regular basis during the employee's assignment, you will want to evaluate the firm's speed of response to your inquiries. If the firm is sluggish in responding to your preliminary marketing call, what can you expect from it when you are a customer? If you have a hard time getting in touch with your account executive before you even have employees on site, you can expect to continue to find it difficult to get his or her attention when you need it.

Reputable Firms As Providers of Reliable Staff

The quality of the firm will directly reflect on its employees. a firm that relates poorly with its prospects and clients is not likely to relate well with its employees. If the firm does a poor job of client service, it is also probably doing a poor job of recruiting, retaining, selecting, training, and placing its employees. The quality of the firm will also impact the loyalty of its staff. Contract employees are sensitive to the quality of the firm they work for. Since they work on an assignment basis, they often do not have the ties of tenure and loyalty that bind a permanent employee to a specific company.

Since many individuals come to contracting as displaced workers, they have been forced to seize their careers and develop a self-interested and sometimes cynical approach. They know that their career is tied to the firm. They will seek better employment with a different firm or company if they do not feel as though the firm is treating them well. A firm's ability to retain its employees is important to its clients.

The stability of the firm's pool of employees will have a direct impact on its ability to fulfill your assignments. The firm that maintains a significant ongoing pool of regular employees will know its employees. The firm will then have a better grip on those elements of the individual's work that do not typically show on a résumé or in an interview. This is very important for "fit" in organizations with highly developed corporate cultures. There is an added benefit to hiring contract staff through firms with stable workforces. The firm's personnel are in fact long-term employees of the staffing the firm. There is less risk for your organization. The employees have been thoroughly screened and have proven their ability to succeed on a contract assignment. These employees will more than likely require less management time. They are better hires.

When you hire a long-term employee from a firm with a stable work-force, you are hiring the elite of contingent workers—those who have chosen this form of employment as a career option. They are less likely to leave for better (read "permanent") jobs and more likely to be satisfied and knowledgeable with the demands and vagaries of the contract work life. These employees know what to expect and have chosen an alternative career path.

As you screen the firms, you will want to carefully examine how they treat their employees. Does the firm provide health benefits for its employees? Do employees accrue vacation time? Do they have access to a retirement plan? Are there provisions for educational advancement? Although someone considering working for a contract firm would consider these issues tantamount, they are also important for the manager choosing a firm to provide staff.

A firm that provides benefits and motivation for excellence to its employees will develop a stable workforce and will be better able to meet its clients' needs during times of tight employment. The contract staffing firm's workforce should not just be individuals earning a wage during a protracted job search. This is very important when you ask the firm to provide individuals with highly selective skills during times of near or full employment when specialized talents may be in short supply.

The more stable the workforce is, the easier it is for the firm to advance its employees' skills and then recover its investment. When the contract employee remains with the firm, it has more opportunity to recoup dollars spent training the employee. When there is the probability that training will be recouped, the firm is more likely to provide this benefit. You in turn as the client will benefit from the contract staffing firm's efforts on behalf of its employees.

The more the staffing firm puts its employees on a par with permanent employees, the more likely you are to obtain a satisfied worker. By providing benefits and other incentives to stay with the staffing firm, the firm can become more selective in recruiting employees. Any organization, whether a contract staffing firm or permanent employer, will find it difficult to develop and manage its staff if their staffing is a revolving door. It is costly and saps productivity to constantly replace workers.

For these reasons, don't hesitate to ask about the stability of the staffing firm's workforce, what benefits it provides its workers, and even how well it pays its employees. You can expect to pay a premium to the firm for the service it provides, but you should also develop an understanding of how the firm pays its employees. The employee who feels (or knows) that he or she is being exploited by the firm may prove difficult to manage or will leave at the first opportunity. Understanding how the firm pays its employees will make it easier for you to work with your account executive in setting a wage rate for any job that you are seeking to fill.

Staffing Firms As Service Businesses

Staffing firms are service businesses. Although many firms approach their business as though staffing were a commodity, the business of technical staffing is more than just providing a commodity. As staffing providers, we must service both the clients we provide staff for *and* the staff placed on assignment. Quality firms not only will hire and retain their employees more readily but they will also offer both parties an attractive level of service. Part of what you should ascertain in evaluating an firm is how it will service both your account and its employees.

The traditional temporary agency would take a job order by phone and send out an individual that in the firm's estimation could fill the position. Unless there was a problem with the individual, the person simply completed the assignment and went on to another. If the person did not work out, the firm would send a replacement. This model assumes readily available replacements and short-term assignments. The longer the assignments, the more important the firm's service becomes. Since most engineering assignments are several months, even years, long, you will want to develop an understanding of how the firm relates to its employees.

You should ask how the firm communicates job assignments and stays in contact with its staff during the assignment. For a long-term assignment, it is not unreasonable for an employee to expect a written description of the assignment. Many firms do not provide their employees with written descriptions of their client assignments. A written description provides the technical hire a clear picture of duties and expectations. It can forestall potential misunderstandings. If the contract staffing firm does not provide this documentation, you will want to. You should share this documentation with the employee and the contract staffing firm. Always remember that the contract staff is the firm's employee. Everybody, however, needs to be on the same page. Do not assume that the job order given to the contract staffing firm automatically fulfills the need for a clear description of the assignment.

You should also find out how the contract staffing firm tests and orients its employees. It is in your interest to get thoroughly screened prospects. You should ask how the firm handles drug testing, verification of citizenship, education, past employment, and references. You should have no doubts about how these are handled. These can create legal complications that you do not want to be involved with. You will also want to know how the firm verifies skills. Does it simply take the individual's word, or does the firm use other methods to verify the individual's qualifications and capabilities? The more carefully the firm screens the employees it sends out, the more likely you are to get the employee you need. This is yet another strong reason to work with a firm that has a strong stable long-term workforce. These firms have had a chance to verify the skills and qualifications of its employees through previous successful contract assignments. The sooner you develop an understanding of how the firm handles this screening, the less you will have to worry about it later.

Employee orientation is another area you should probe. Although you will want to provide the engineer with an orientation specific to your organization, you will want to know what sort of orientation the firm offers its employees. Firms that provide benefits and support for their employees usually provide some orientation program for the employee. Orientation makes an employee feel as though he or she is part of an organization. The orientation may be as simple as a quick briefing about benefits and their eligibility, or it may be more extensive. Some firms provide their employees a handbook and go to considerable effort to make sure that the employee feels the support and commitment of the firm before being sent out on assignment. When employees feel cared about by their employer, they develop loyalty. This leads to worker satisfaction and will in turn benefit you as a client.

Not only should you ascertain how the firm handles the preemployment phase, you should also ask what contact the firm will have with the employee during the assignment. You should determine whether the paycheck will be the only contact with the firm. You will want to know how closely the firm will monitor with you the employee's performance during the assignment. Although it is your intent to hire a worry-free employee, you should know in advance how the firm handles any problems that might arise. Whom would you contact if you had a disagreement with the employee or had issues on performance or qualifications? What are the firm's policies, and how does it handle employees that don't work out? The more you know about how the firm will manage its employees—your contract engineers—the more secure you will become about how you will work with the firm.

You should also ask about what kind of training opportunities the firm provides its employees. Some firms provide very little ongoing training for their employees; others offer tuition reimbursement for courses taken at colleges and universities. The more relevant training the firm provides on a long-term basis, the more apt their employees are to have current skills. In rapidly moving technical fields such as information systems, communications, and most engineering specialties, employees must be constantly learning or their skills will erode. Although the contract employee will learn something on every assignment, you should look for the firm that does not just rely on skills learned on the job for employee training. The firm that trains its employees will provide you employees with enhanced and current skills.

Technical employees also recognize the value of training as a key to a higher wage rate and more interesting assignments. They will more readily develop a long-term attachment to a firm that provides these growth opportunities than one that does not. As the client of the firm, you will find the employees happier with their work assignment and easier to manage.

Since the employee will remain with your organization for a substantial time, you will want to ascertain how the firm handles employee performance appraisal. Is the employee evaluated annually, at the end of each assignment, or on some other schedule? Services firms with long-term employees usually have performance appraisal systems. The contractor has the most contact with the employee and is most able to provide feedback on the employee's performance. It is the firm's responsibility to evaluate its own employees. As the contractor, you will be expected to provide information for this evaluation. You should know before you contract with the firm how the appraisal process works.

Some firms provide more continuous feedback to their employees. Others give very limited information. When the firm provides ongoing support and coaching for its employees, your job is easier. The firm with its ongoing contact with the employee in essence is contributing managerial support. Since you will be communicating your assessment of the contract employee's performance to your firm contact, you will again find it beneficial if you can build a solid working relationship with your account representative.

Another area that you will want to explore is how the firm handles converting temporary into permanent hires. Contract staff can be a valuable source for recruiting permanent staff. When you have had the opportunity to not only review the résumé and interview the individual but also observe their work habits and organizational fit, you may decide that the contract employee is the ideal candidate for a permanent position. Before this situation arises, you will want to know what type of arrangements the firm makes for converting employees from contract to permanent. Evaluate in advance the costs and procedures. Many firms such as IBM, Ford Motor, Xerox, and Merck use temporary assignments as trial hires for prospective permanent employees. Contract to permanent hiring reduces risk and costs, but you should inquire in advance.

Finally, if your firm has ISO 900X certification, you will want to use firms that have sought this certification for themselves. The firm is a provider of inputs (labor) for your manufacturing processes. The firm will meet the requirements as a supplier. Also, the firm will have a real-time understanding of the requirements as a result of going through the certification process and the requirements it places on staff. This will in turn transfer to its understanding of your needs.

When you are evaluating technical staffing firms, you will also want to ask if they are members of local business organizations and whether the firm holds membership in the National Technical Services Association (NTSA) or other industry organization. Members of NTSA must meet specific requirements and ascribe to a code of ethics that supports ethical conduct toward the technical staffing industry, its employees, and the clients it serves and the public. Not only must members agree to adhere to the organization's code of ethics but risk censure and expulsion for breaches. When you deal with firms that are members of business and professional organizations, you are assured of their commitment to the community and industry.

Making Your Choice

After you have evaluated the qualifications of each firm and are ready to begin hiring contract staff, you may find it helpful to submit the same job order to several firms that meet your approval. You can then begin a real-time evaluation of their service and the employees they provide. It is not uncommon for an organization to use multiple staffing firms to provide differing types of employees. For example, clerical employees may be supplied from a different firm than CAD operators, designers, or IT professionals.

If you decide to use several firms concurrently, you will need to recognize that there is an inherent cost. The cost is in your time and in handling any paperwork or other administrative activities. Some companies that rely heavily on contract personnel have found it advantageous to develop in-depth relationships—strategic alliances—with a single key provider. In Chapter 8 (pgs. 159-163), we will discuss the advantages and disadvantages of these relationships and how to forge them.

Summary of Key Points
in This Chapter

1. Don't wait until you must hire under pressure to begin evaluating available contract staffing firms.

2. Before selecting a contract staffing firm, seek advice from human resources and peers already using contract staff.

3. Your ability to get the employees you need depends on the firm, the job description, and the labor market. You control the firm choice and the job description. Seek to mitigate the impact of each element.

4. A successful client-firm relationship requires trust, rapport, and effective communication between the account executive and the contractor.

5. Seek firms that service your industry and geographic area.

6. Evaluate the firm's responsiveness, recruiting practices, screening processes, and management of employees on contract assignments.

7. Seek to hire contract employees with tenure and experience in contract work.

8. Look for firms that provide career incentives to their employees to help them build stable workforces.

9. Make sure you understand before contracting how the firm will handle assignments, problem employees, evaluation, and shifts in employee status from contract to permanent.

10. If your organization is ISO 900X certified, look for firms similarly certified for your contract labor suppliers.

11. Although it is not uncommon for organizations to use the services of multiple staffing firms, consider developing a strategic supplier relationship with a firm of choice to reduce process time and cost.

8

Developing a Strategic Alliance with a Contract Staffing Provider

Marge is the human resource manager at a third-tier automotive supplier. Her company manufactures large volumes of engineered parts. Downsized 3 years ago, her department must handle all human resource functions for a large manufacturing facility and the engineering and design department. One of the positions eliminated in the downsizing was the department's technical recruiter. For the past year, Marge has struggled with staffing the design department to meet a current dramatic upswing in demand. The design department is now larger than it was at the time of the downsizing. Given the cyclical nature of the automotive industry, management has refused to add permanent design staff and has not authorized Marge to rehire the technical recruiter. To meet the increased technical staff hiring needs of the design and other departments, Marge has relied heavily on contract staff provided through several different staffing firms. She is very pleased with the results; however, managing multiple contract vendors is taking a lot of her time. She has considered choosing to use just one of the contract staffing firms exclusively. She knows that she should discuss this with the managers who use the contract employees, but

*who else should be involved in making the decision? She
has jotted down a list of questions that she will need
answers to: What sort of relationship should she seek to set
up? What is possible? Which firm could provide the best
support? How will she get the mix of staff she relies on
from several staffing firms? She looks at her list and tries to
think of what might be missing.*

As corporations try to become nimble competitors in a tough world
market, they must look for ways to reduce costs and wasted effort
throughout the organization. Staffing and human resources are often
cost-reduction targets. Cost reduction in business has focused on the
economic inputs and streamlining the processes that transform raw
materials to finished goods ready for the consumer's use. Economists
have long considered labor as an input along with raw materials, capital, and enterprise.

American business has clung, however, for many years to a labor
model that did not provide for rapid adjustments in volume. When manufacturers moved from stable workforces to using more contingent labor-
ers, they developed a new model for managing this element of the economic mix. Labor, particularly in engineering and other overhead areas,
was traditionally a more static input than raw materials in the manufacturing process. However, labor is a key to transforming raw materials into
finished goods and should logically be as flexible as raw materials.

In manufacturing areas, adjustments in labor demands were met traditionally with layoffs and overtime. At best, this is an imperfect
method of adjustment. This simply stretched or shrank the output of a
static staff. By using contingent and contract personnel, manufacturers
today can adjust the output, the volume, and the skill level of their
workforce and thus fine-tune the labor component of the economic mix.

In the past, companies less pinched by competition had the luxury of
keeping engineering and technical talent on staff although underuti-
lized during slack periods. The assumption was that the company
needed to have the talent available for future projects. Underutilization
was simply a cost of doing business. These research and development
and design departments were expensive to maintain. As competition
heated up and profitability was challenged, they became targets for
reduction and elimination.

Since the need for research and development and design did not go
away, new staffing patterns for these areas have emerged. Today, many
research and development departments have a minimal permanent staff

supported by contract engineers and technicians chosen because they have skills that specifically match current project needs.

Traditional staffing models are vanishing. Today's forward-thinking companies have developed strategies that allow them to flex their technical workforces just as they might any other labor input. It is not surprising that many of the same contractual relationships that are used for handling raw materials and equipment are now being applied to staffing. The development of strategic alliances with contract staffing firms simply parallels other corporate shifts to interdependent working relationships that reduce cost.

The Value of Interdependence

There is a growing recognition of the strategic value of interdependencies. These initially focused on the supply and distribution chains. By entering into relationships that reduced cost or improved service from suppliers and to distributors along the chain, manufacturers could better position themselves to respond to the marketplace. A strong supply-distribution chain has become a key to competitive advantage.

This shift, however, required a rethinking of supplier-distributor relations. Over the years, the need to maintain adequate supplies of the thousands of parts and components that go into complex manufactured products such as automobiles, aircraft, and electronics had created a maze filled with thousands of specialized suppliers. Recently, these large, clumsy networks of suppliers have themselves become targets for cost cutting.

Today, there is both recognition of and intolerance for the once hidden costs of supplier relationships. For every part, one must consider not only the cost of the product itself but also the cost of specifying, prototyping, and developing the part and for verifying quality, handling the paperwork, and maintaining the vendor relationship. When the number of suppliers bulged into the thousands, companies found that the cost and problems with managing these relationships created a considerable drag on organizational agility.

Management is now focusing on reducing the number of suppliers and increasing the strategic value of each relationship. Every supplier must now add value and increase the company's ability to compete. Since almost every purchaser is a supplier to another from raw materials to finished goods, there are many interdependent links in the chain. For companies to gain strategic strength, each link must add value for the next. The amount of value added determines the strength of the chain, and each link will enhance the competitive advantage of each organization in the chain.

The ability to add value and increase competitive advantage are the measures for supplier evaluation. Value-added suppliers are expected to do more than just provide an input into the process; they must add to the company's ability to compete effectively. This is true whether the supplier is providing raw materials, finished goods, or contract staff.

An example of this type of enhanced, value-added interdependence is the just-in-time (JIT) supplier. By providing a steady stream of inputs to the manufacturing process to mesh with the manufacturing plan, the company can reduce its need to maintain parts inventory. A parts inventory is considered "fresh-frozen assets." Inventories are costly, potentially perishable entities. They must be tracked and cared for. Any engineering or model change can render the inventory obsolete. Most significantly, the assets tied up in parts inventory are not available for other corporate needs. They are *frozen* in the warehouse. JIT, by providing only the parts needed for current manufacturing, reduces the need to tie up assets in parts inventory. The JIT supplier adds value by saving the cost of maintaining the inventory and providing a reliable stream of parts.

The relationship is usually structured to also provide advantages for the supplier. These are framed around guaranteed volumes, attractive financial results, and opportunities for innovation. The details of the working arrangement are spelled out contractually and depend on the supplier and the manufacturer.

Establishing the relationships behind JIT manufacturing scenarios required rethinking traditional supplier and vendor-purchaser relationships. In the past, each party worked from keen self-interest with near-total disregard for the interests of the other. The supplier was eager to get the order, and the buyer was looking for the best value. Value was often framed entirely in price. Mutual interest and cooperation were limited.

The shift from traditional supplier relationships to JIT did not occur overnight. This shift represents a sea change in supplier relationships. It required an understanding of the interdependence of each entity and a willingness to recognize the strategic advantages of mutuality. To make a JIT relationship work, both the buyer and the purchaser in effect have had to lay aside a singular focus on their own self-interest and consider the interests of the other party and how they might affect one another. When win-lose thinking is replaced with a win-win perspective, one sees the potential value of cooperation and mutuality. For these relationships to work, there has to be significant discernible advantage for both parties.

In manufacturing scenarios, the relationship is reasonably obvious. The supplier gains a customer with an established working relationship and dependable volumes; the purchaser eliminates inventory needs. Interdependent supply relationships are not limited to raw materials, but they can be developed to suit any of the economic inputs including labor. A strategic alliance with a contract staffing firm can provide a relationship with significant mutual advantages (Exhibit 8-1). If you decide to develop a strategic alliance to meet your contract staffing needs, you can be sure that the contract staffing firm will expect mutual benefit in the arrangement. To come to contract, there will have to be advantages for both parties.

To set up these relationships, both organizations must develop not only trust but also strong communication links. There must also be a recognition of the financial interdependence of the two organizations and an awareness of the impact of the contract on both organizations. When manufacturers use JIT, they give up some control over their sup-

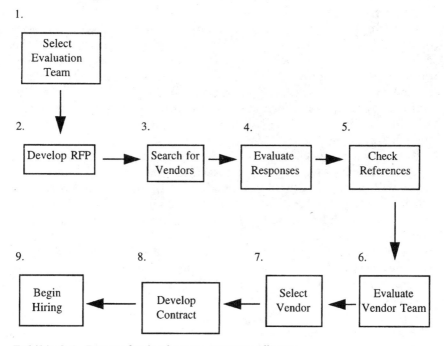

Exhibit 8-1. Process for developing a strategic alliance.

ply chain. They no longer have the warehouse of fresh-frozen assets—inventory—to fall back on when the supplier does not meet demand. Similarly, a contract staffing firm in a strategic relationship must also be able to reduce the customer's reliance on staffing "inventory."

In manufacturing, when anticipated inputs are unavailable for the manufacturing process, the manufacturing process will halt. This can potentially create financial loss. The same dynamics apply to staffing. When demand for staff increases and contract replacements are unavailable, projects are delayed, and windows of opportunity may close.

The recognition of the fragility of the relationship and the potential for downside losses has made management cautious in setting up heavily interdependent relationships. In some instances companies seeking to establish interdependent, economically positive relationships have preferred to use joint ventures or to completely outsource departments that are not strategically aligned with the core business. In these, both parties share the risks and rewards of a new venture usually remote from the core business. The risks are often greater in a JIT manufacturing relationship because the supply chain is closer to and has greater impact on the core business. When you consider setting up a strategic alliance for contract staffing, be sure to consider the impact of failure and success on the core business.

Up to this point we have focused on inputs. However, a ready supply of inputs provides just one potential source of advantage. To increase efficiency and improve operating performance, companies are focusing on improving operational processes, for these are also at the heart of competitiveness. By applying process improvement techniques to staffing, companies can increase performance and achieve competitive advantage. Changes in the handling of contingent labor can provide these advantages.

A strategic alliance with a single contract labor vendor should be structured to improve the labor contracting processes. For example, the labor contractor can provide a single bill, albeit very detailed, that will cover all of the contract employees hired. This will reduce the handling in the finance and accounting departments. Organizations with large contingent workforces have found that just managing these workforces represents a significant human resources drain. To offset this effect, staffing firms will provide an on-site supervisor to handle a variety of functions related to the management of the contingent workforce.

When there is a representative of the staffing firm on the premises, the contractor is able to relate to the firm directly, reducing managerial time. These relationships can also streamline the hiring process since the connection is more direct. The supervisor becomes responsible for ensuring that there is adequate staff on hand to meet changing needs. The on-site supervisor from the staffing firm becomes a vital link in the production planning process. For organizations with large numbers of contract staff, the staffing firm's representative handles many other human resource responsibilities such as orientation, training, and deployment to work assignments. Where there are significant numbers of employees, the result is less burden on the contractor's human resource function. Since the individual on site is a specialist in dealing with contract employees, the employees have a direct and informed source to help meet their needs. This can result in a more satisfied workforce with lower turnover and absenteeism.

Seeking an Alliance for Contract Engineering Staffing

The growth of contingent and part-time labor suggests an understanding of the potential for handling staffing in a JIT mode. Contract engineering and technical staff, however, are subset of the contingent labor force that do not fit the model in the same way as manufacturing staff. With their broad array of highly specialized skills, they are best deployed on long-term projects. They are not the same as the contingent staff hired for low-skill assembly jobs on an almost daily basis.

In building a strategic relationship with an staffing firm for contract engineers, you use a slightly different focus than you would for developing a staffing contract to provide contingent staff to meet manufacturing needs. The engineering and technical staffing firm needs to know more about the forward planning of your business and the technology that you intend to apply over time. This extends beyond production planning to new-product and new-technology planning. The firm will be your partner in the long-term management of contingent technical employees and needs to be planning for how it can meet your needs, not just now but in the future. This means that you will need to develop a number of decision criteria for selecting your strategic ally that hinge on this long-term focus.

In the last chapter, we examined how to select a contract staffing firm. The process and the criteria for selecting a strategic staffing ally will go beyond those used for selecting a firm to provide a few employees to a single location on an ad hoc basis. In addition, it should follow the process used for setting up other contractual relationship. Exhibit 8-1 shows the steps that should be involved:

1. Select the evaluation team.
2. Develop an RFP.
3. Search for vendors.
4. Evaluate their responses.
5. Check references.
6. Evaluate the vendor team.
7. Select the vendor.
8. Develop the contract.
9. Begin hiring and implementing the contract.

The contract itself is often negotiated at the corporate level and is the result of a series of meetings and inputs. A strategic staffing alliance is not a trivial undertaking. It may have broad impacts on staffing across a number of departments and locations. It is important that the project team set up to evaluate the prospective vendors represents as many different areas and departments as will be impacted by the decision. It is not a single-person decision; however, the size of the project team will depend on the decision-making culture of your organization. Some companies, although using a team structure of establishing decision criteria, will embody a single individual with the responsibility of serving the needs of the entire group. No matter how the decision is made, it should reflect the needs and wants of all of the impacted parties.

Once the composition of the team is decided, it is easier to determine the criteria for selection and to develop the structure that will support the decision-making process. The formal process for communicating and gathering vendor information can be established in a manner that will benefit all parties. There may be a need for some preeducation on the scope of services available through a staffing agency. This fact-finding should be done before the RFP is let. This will enable your decision group to clearly articulate your needs to the companies vying for your business. Although the decision criteria established by your group will reflect your own needs, the following are some factors that you will want to consider in evaluating the responses you receive:

Their size versus your size. Any strategic alliance must partner your organization with one able to meet your needs. Size is often a key determinant of the firm's ability to accomplish this. A small firm operating out of a single centralized office may find it difficult to supply staff to multiple plants scattered across three or four states. If you are looking for a staffing provider to meet your needs across a wide geographic area, you will want to find one that has the size and office placement suitable for your needs.

If you expect to use the firm to provide a significant number of contract technical staff almost immediately, you will want to work with a staffing firm that already provides numbers of similar staff to other similar businesses. You do not want your contract to instantly multiply the firm's billings by an unworkable factor.

Their strengths versus your needs. Your chief need should be the firm's strength. Many contract technical agencies specialize in the types of staff they provide and industries served. Some specialize in data processing professionals, others design engineers for automotive or aerospace. The recruiting needs of these specialized industries are so specific that you will want to develop your alliance with a firm whose strength is in your primary need area. If you anticipate using the alliance to fulfill a need for CAD designers, look for a firm that provides large numbers of CAD designers. This does not mean that you should eliminate a firm from consideration just because they cannot provide *all* of the staff specialties that you might need. Your strategic ally can subcontract with other specialty agencies to fill the gaps.

Their network and your needs. To enable the firm to recruit across a broad band of technical specialties, you may want to structure your arrangement to permit the firm to secure engineering talent through subcontracts with other agencies. In this instance your strategic ally will function as a general contractor. If you anticipate this type of relationship, you will want to evaluate whether the firm has the ability to accomplish this. Is their network large enough to furnish the contacts needed to fulfill this type of contract? Do they have the stature and reputation that will enable them to secure quality employees, or will the firm always be scrambling to get subcontracted help? Assessing reputation and stature is difficult; however, your preliminary screening for a contract staffing firm through your contacts and professional network will yield surprising results. Reputations are made and lost at the hands of customers and their comments.

Your experience with them and their response. Before you consider developing a strategic alliance with a contract staffing firm, we would recommend that you hire contract employees through the firm. In letting your RFP, you should strongly consider those firms with which you have already established some type of working relationship. This can provide valuable insight into how the relationship might work in the future. This may not always be possible, but it does provide greater assurance of whether the service will mesh with your stated needs.

When you have a working experience with a staffing firm, you can develop a profile of how a firm responds to your needs and fits with your business approach. This trial period will give you a chance to determine how the contract staffing firm recruits, selects, hires, orients, and manages its staff. Be sure to check whether the contract staffing firm selects and recruits staff of the quality and skill level that you need. Are the employees satisfied and easy to manage? You do not want a growing number of contract employees to sit on your organization like an indigestible meal of malcontents.

You will want to verify whether the staffing firm provides engineers and technicians whose skills and performance match your job orders. Do you find that you have to try several potential hires before you are satisfied with those presented by the staffing firm? When multiplied by larger numbers, this could require significant management time and result in efficiency losses. Check how responsive the contract staffing firm is to your needs. If the contract staffing firm is going to be filling a large number of job orders for you, you will not want to wait while it seeks the individuals to meet the job orders.

Although you checked into how the firm handles orientation before using it, now review how the firm actually oriented the employees it sent. Orientation is very important when the firm will be providing numbers of employees. If you are unhappy with how the firm prepares a few employees, you will really dislike it when the numbers swell. You may want to include details on the orientation process in the contract.

Since you will be using the firm to provide a number of employees, you should also check how it handles the paperwork. Do you find yourself disputing or having to clarify the billing on a regular basis? Are the firm's support personnel helpful and cooperative, or is the account executive your lifeline in an otherwise tempestuous sea? Since one of the reasons for using contract staff is to reduce managerial input and streamline processes, you will not want to spend endless hours chasing after answers to the simplest questions and playing telephone tag with an unresponsive firm.

When you have not had the opportunity to work in real time with the vendors responding to your RFP, you will find it doubly important to check the references and to determine how the firm fulfilled its role. Just as with individuals, firms choose references that have had a positive or above average experience with them. Do be sure to probe for any difficulties that might have come up during a reference customer's experience with the firm. Then, consider how the situation was resolved. The resolution of problems to the satisfaction of the client is often the mark of a good working relationship. It shows the organization's capacity to respond to difficulties and work them through to mutual satisfaction.

Prior to selecting a vendor, it is important to evaluate the vendor's team and the infrastructure that will support your contract. This should be done prior to formalization of the vendor decision in a contract. Just as in selecting a firm to provide a few contract employees at a single location, you will want to determine how well you can communicate with the account executive. If you have difficulties or discomfort in dealing with the account executive or firm management, developing an alliance will be virtually impossible. Since the relationship closely ties your organization to the firm, you will want to ask yourself, "Can I trust these people?" If you do not think that you can, consider another firm. The sense of discomfort will color all the negotiations and ultimately the working relationship.

Negotiating the Contract: The Provisions and Considerations

Once you have chosen a suitable partner and determined that the firm is willing to work with you, you will want to negotiate a contract to cement the relationship and to detail key points in its operation. The terms of the contract should of course be subject to legal approval, but here are some of the points that you can expect to negotiate.

Which of the organization's needs will the contract cover? You will want to specify which units or departments will be covered by the arrangement. If you have a large organization, you may establish a strategic staffing alliance to cover just design or data processing contract engineering needs. On the other hand, you may find it advantageous to set up arrangements with one firm for securing specific types of contract engineers for multiple units. For example, you may want the firm to secure all of the data processing contract employees for units across the organization, whether the operations are multistate or multiunit in the same geographic area.

What jobs will the firm fill and at what billing rate? Perhaps the most important point of negotiation will be the specific jobs that the firm will provide and their respective billing rates. Before you submit the RFP, you will want to review your current pay rates for those jobs that you expect to hire through the firm. Develop a list of titles, matching job descriptions or summaries, and salary ranges. Although the job descriptions may already be in your standard format, you will need to review them. Take a careful look at the qualifications and skills you will require of each job. Since you will be hiring contract personnel whose experience and qualifications are the basis on which you will evaluate them, be sure that your job descriptions clearly spell these out. For more details on developing job orders for contract personnel, see the next chapter.

You do not want to either underpay or overpay the firm personnel. You should, therefore, compare not only each firm's charges to you for comparable jobs but also the direct labor rate each pays its staff and the benefits provided. Although you may gain an economic advantage by obtaining personnel at significantly less than your standard pay rate for the same job, you should be aware that this strategy may backfire. Two-tier workforces are often a source of strife and lead to discontent among the permanent workforce. The concern of permanent staff is, of course, that you will either eliminate their jobs and refill them with lower-paid contract employees or seek to convert the current staff into lower-paid contract employees. The perception of pay inequity will make it particularly difficult to shift employees from permanent to contract if this is a strategy you want to employ. This perception can also reduce your ability to flex your staffing to meet long-term strategic needs.

What provisions are there for subcontracting? You will want to consider the impacts of an arrangement where your strategic ally acts as the key contractor for all of your contract staffing needs. You will want to be sure to spell out any subcontracting provisions both in the RFP and in the final contract. These arrangements can give you access to a broader network of potential talent; however, they also add another link to the process. You will want to be sure that the details of the relationship are clearly articulated. It is not uncommon for staffing firms to work in conjunction with one another. This allows the general contracting firm to obtain staff in disciplines that it does not necessarily specialize in.

For example, a firm that provides large numbers of CAD designers may be called upon to provide quality-assurance technicians. The firm may seek these staff through another agency that handles more work in this area. When this occurs, the employees work at two steps from your organization. They are employees of one firm working for another at your site. You will want to be sure that you understand who the employee actually works for and how he or she will be managed. You may even want to have the right to eliminate some potential subcontractors from consideration based on previous knowledge. You will want to air these issues during the negotiations so that you do not tie your staffing ally's hands, preventing it from providing the needed staff.

What work rules and policies must the contract employees follow? When permanent employees arrive for work, contract employees can be expected to arrive at the same time and work the same standard hours under the same conditions. However, your contract needs to address workplace-, or company-specific deviations from these norms. For example, if the employer gives employees the option of working four 10-hour days, will contractors follow suit? What happens when inclement weather closes a plant or makes road conditions too hazardous for safe travel? What do the contract employees do when the permanent employees take Patriot's Day off to watch the Boston Marathon?

On a less pleasant note, what is your expectation of the contract employees during a labor dispute? Do you expect them to cross picket lines? Do you expect them to show up during a lockout? What options do they have if they feel threatened? As you consider these issues, you should be up front with your potential allies as to the possibility of a labor dispute at the site you expect the firm to provide staff for. The vendor will want to consider this problem in responding to your RFP. Then thoughtful consideration of these possibilities during contract negotiations can greatly simplify communication at times when you, the contract employer, will be preoccupied with other concerns.

Who bears responsibility for orientation and training? You should spell out who is responsible for the orientation of *all* contract personnel, including subcontracted personnel. One of the advantages of setting up a strategic alliance is the opportunity to get the staffing supplier to share the responsibility and costs of orientation and training. When the number of contract employees reaches a critical mass, it makes sense and is cost-effective for the firm to handle orientation. When this occurs, however, you again lose some control. Clear expectations effectively communicated are key to mutual satisfaction.

Similarly, arrangements should be stated as to who bears responsibility and costs for ongoing training of long-term technical personnel. Staying competitive requires continuous investment in the human capital of the firm. If there are no provisions for continuously training both your permanent and contract workforces, you are potentially hurting your competitive position.

You may even decide to include contractually where the orientation and training is to take place—on your site or at a site provided by the firm. It may behoove you to combine training efforts. It is just as easy to run a training class for 20 workers as for 10. Then, the issue would need to be worked out as to who provides the training and how it is billed back to the other partner. Shared training expenses may actually result in cheaper training for both companies.

What is the cost and process for transferring employees between the organizations? Just as you will want to spell out how contract employees are oriented to your business, you will want to identify up front any provisions you wish for transferring employees from the firm's payroll to your own and the converse. For example, you may wish to determine how you can reduce your permanent payroll through transfers to the staffing firm, payrolling (a popular option for organizations undergoing reengineering and outsourcing). You will want to determine not only the mechanics but also the costs for these shifts. Your contract may include provisions for using interfirm transfers for outplacement. For example, a downsized engineer could be transitioned to the firm for a specified period of time and then become available for other assignments with the contract staffing firm. This would provide the engineer a soft landing in the downsizing.

A transfer to the contract firm is often an ideal arrangement for an employee wishing to wind down a career—the retiree who is not quite ready to call it quits. You can retire the individual and then use his or her services on a contract basis through the firm.

You will want to include provisions for converting a contract employee to permanent status. Many contracting agencies serve as primary recruiters for engineering staff. This role transfers to the staffing firm the cost of recruiting and orienting the employee and lets you then evaluate the prospect on a real-time basis. You cannot expect this value-added service to come for free. Your contract should state the costs and the process.

How long will the contract run, and what are the provisions for extension and termination and change? Although you will enter into and negotiate the contract to cover a specific time period—1 year, 2 years, etc.—you will want to be sure to include provisions that will allow you to extend, terminate or change the contract. Beware of the lengthy contract. Many heralded long-term contracts burn brightly for a while and then become increasingly difficult to maintain. Their disintegration can become a source of embarrassment for both firms. You are wiser to include realistic contract periods and provisions that will protect you in the event that you become dissatisfied with the firm and want to end or alter the relationship.

With mergers and acquisitions an ever-present possibility, you will also want to include provisions for changes in control—both yours and the firm's. You will want to spell out how such actions could impact your arrangement. Similarly, you will want to determine in advance how you will handle plant closings, whether permanent or resulting from long-term labor disputes.

What processes and infrastructure will support the contract? Whether the firm will be providing staff to one or a number of different sites, you will want to establish how the billing will be handled. Will you want a consolidated bill for all units? Where will it be sent? What are the payment discounts? These can add up, if you are using a large number of employees.

What is the procedure for submitting and filling job orders? Where are they to be submitted? You and the firm will want to agree on the central contacts within each organization and make sure that all parties understand and are prepared to fulfill their roles. It is important to set up the infrastructure that will support the contract before you ink the final deal. This is why evaluation of the vendor team precedes the selection of the vendor in the process outlined in Exhibit 8-1.

Getting the Support You Need from the Firm

Reducing the human resource management workload is a key factor in deciding to establish a contract staffing alliance. It is important to structure the relationship so that it yields these results. One of the ways this can occur is to negotiate that the firm maintain a presence in your organization. With a vendor-on-premises arrangement, the staffing firm assigns a staff member to directly manage the contract from within your walls. You either allocate space for the individual, or the firm rents it from you. This is similar to corporate travel agencies, where the travel arrangers are actually employees of a travel agent. They are on premises for the convenience of the client. The travel firm maintains an office (often a cost to them) in the company's headquarters. This gives the agency unparalleled access to the bookings and client the convenience of in-house travel arrangements.

With a contract staffing alliance, the vendor on premises can enhance the communication between you and the firm as well as provide human resources support for your contract staff. This benefits you and the contract staff. Since the firm is the employer of record for your contract workforce, the firm has the responsibility to provide them access to the firm's human resources. It is through the firm that they receive their paychecks, benefits, vacation, transfers to other jobs, etc. The employee must contact the firm for this support. By having a human resource specialist affiliated with the firm on hand, the employee can receive answers to questions and complaints without having to leave the job site and with minimal effort. This will increase employee satisfaction and improve the employee's productivity.

The presence of the firm on the job site can increase the employee's loyalty to the firm. There is a visible, human connection point at their job site. This will benefit the firm in the short run and you in the long run. Since contract technical staff are usually on long-term assignment, there is an increased probability that the employee will need to contact the firm about a human resources matter at some point during the contract than an employee working for a few days on assignment. Where there are numbers of long-term contract employees, this issue takes on larger proportions.

You, the contractor, can reap the most benefit from this arrangement. Not only does the firm have direct contact with its employees but you have access to the firm on demand when you need it. As we previously indicated your on-site supervisor is key in helping maintain adequate staffing to meet shifting production needs. Having the supervisor at hand will facilitate discussions on hiring of additional staff, the interviewing for prospective new hires, and the orientation process as well as the routine management of the contract staff. When the staffing arrangement has brought in personnel from several agencies on a subcontract basis, you will find it helpful to have the firm's specialist serve as a hub supporting both you and the contract staff.

The greatest benefit, however, is when problems arise related to contract employees. Whether the problem is a safety violation, an injury, a supervisor-employee disagreement, or a sexual harassment claim, the firm is immediately available. When you want to transfer or end an employee's assignment, the firm's personnel are there to facilitate the transition. The firm representative will also keep track of the status of each contract employee and serve as a go-between to ensure that those near the end of their contracts are appropriately transitioned either to another assignment within your organization or to another contract.

Determining whether your contract is large enough for the firm to economically maintain someone on premises on a full-time basis is part of the negotiations, and your expectations should be stated in your RFP. As you begin looking for a staffing ally, you should consider whether you anticipate having enough contract staff you to command this level of support. Its feasibility will depend on the billings you generate for the firm. If you and the firm do not determine that the contract merits a full-time contract manager, you should negotiate to have the firm personnel on hand at regular intervals to handle contract employee human resource issues.

Since some agencies prefer to deliver their employees' paychecks on site, you may find that you want the firm personnel to stay on site for several additional hours or arrange a convenient time. Part of the logistics will be to arrange a location, usually within your human resource department, for this. Since the firm will want to notify its employees where and when their personnel will be on hand, you will want to establish a regular time and place.

You will want to ensure during your contract negotiations that you have the right to approve the firm contact assigned to your organization. The underlying basis for a successful staffing alliance is communication and trust. If you (or the contract staff) cannot communicate effectively with the specialist assigned to your organization, you will not get the support that you need. You will also find that if you (or the contract staff) do not trust the individual, the relationship will not work.

Although we have tried in this chapter to outline some of the criteria for developing a successful staffing alliance and some of the issues that you might consider, we would like to stress that contract staffing is an evolving industry. The relationships and the details of the contracting process are not carved in stone. Staffing agencies, such as ITS Technologies, are in the business of creative staffing and frequently welcome participating in arrangements that provide mutual benefit for the contractor, the firm, and the engineer.

Summary of Key Points in This Chapter

1. Labor is now recognized as a flexible economic input and is treated as such.

2. Contract arrangements to provide contract staff mirror those for other economic inputs such as raw materials.

3. Strong interdependent supply and distribution chains lead to market advantage.

4. Before developing a strategic alliance with a contract staffing firm, look for one with compatible size and strengths.

5. Review the firm's reputation and access to technical specialties through other agencies.

6. Before developing a relationship, use the firm and evaluate its performance during the trial.

7. Negotiate the departments (sites) to be served, jobs to be filled and billing rates, subcontracting provisions, work rules and policies, orientation or training procedures and responsibilities, transfer costs

and processes, provisions for contract extension and termination, and any needed support infrastructure.

8. Include in the contract the level of firm on-site support required.

9. The contract staffing industry is dynamic and evolving, so seek to develop creative staffing alliances that will meet your changing needs.

9

How to Get the Contract Employee You Need

"So how are you and your contract engineers getting along these days?" Morris asked as he slid his tray onto the table where Jack was sitting. From Jack's frown, Morris knew immediately that he had asked the wrong question. Jack began a long monologue about how he just couldn't get the right people. The engineers the staffing firm kept sending were nice enough, but they just didn't seem to have what it took to work in his department. Jack went on to describe all the steps he had taken based on his previous discussion with Morris. Yes, he was using the same firm that Morris used—the one recommended by the company. As Morris probed further, he discovered that Jack was even working with the same account executive. Jack seemed frustrated with the firm and the account executive and blamed them. He complained, "The time I've spent trying to get this position filled is time I don't have. I'm just getting further and further behind."

Morris had never had such problems, and he was puzzled at Jack's difficulties. The staffing firm always had rapidly filled his job orders and sent him qualified people that worked out quite nicely. His department was not that different from Jack's. Maybe the problem was something that Jack was doing? What was he asking for on the job order? Was it a problem of communication with the account executive? Was there a really problem with the staffing firm?

Selecting a top-notch staffing firm does not guarantee a manager successful recruitment of contract staff. Consider how many golfers own the best
sets of graphite-shafted, titanium-headed clubs and still slice the ball, miss
the fairway, and tour every bunker. Selecting the best firm for your needs
is only a first step. It gives you the tools for success, but you must use them
properly. The second and equally important step is communicating to the
firm as precisely as possible the qualifications of the employee you need.
This will ensure recruitment with minimal effort. Just as good mechanics
will improve a golf swing, there are steps you can take that will help you
more rapidly get exactly the employees you need. Hopefully, if you follow
these steps, you will have Morris's, not Jack's, experience with contract
staffing. If you are already hiring contract staff through a staffing firm, you
will find in this chapter some tips to speed your future hiring and help
ensure that you get the right person for every job.

Every Candidate Is Equal
to the Sum of the Parts:
Training and Experience

Every employee, and thus every candidate for a contract or permanent
job, is a unique accumulation of training and experience. No two individuals, even given identical educations and careers, are the same.
Because of our differing personalities, we all process information and
learn differently. This is one of the reasons it is so difficult to evaluate a
candidate's résumé.

You can expect a recent college graduate with a degree in electrical
engineering to have a specific core of knowledge and competencies in
the field. Similarly, an engineer with a PE license would also reflect a
tested level of engineering competency. Assessing knowledge learned
through course work or licensed through a regulatory agency or certified
by an industry-governing body is straightforward compared to evaluating what the person has accumulated as a result of his or her work experience.

Unfortunately, there are all too many individuals whose résumés list
10 years of experience who in reality have had one year's experience
repeated 10 times. Because of the difficultly in evaluating on-the-job
learning, employers have relied on references and interviews and sometimes testing to probe a prospect's experience and personality. In a contract employment situation, the contractor depends on the individual to
immediately function at a high level due to his or her educational background and accumulated experience.

Since the contract employee is not expected to remain with the company and develop the way a new college graduate might through working on an entry-level permanent job, the contractor must obtain individuals who can immediately function at an anticipated level. There is often no time for training and learning unless the hiring organization has proprietary processes, systems or equipment that preclude previous experience. The contracting company has neither plan nor time to invest in the engineer's development. Since the staffing firm is providing the employee, you must be able to communicate to the account executive exactly the level you expect the employee to function at upon arrival on the site.

A common complaint of dissatisfied contractors is that there is a gap between the skills they need and the individual sent by the staffing firm. A reputable staffing firm will provide you the greatest opportunity for obtaining qualified people. If you do not communicate exactly the qualifications you will need, even the best firm will have difficulty satisfactorily filling the position. All jobs have specific duties and attendant tasks that the individual is expected to perform. Education and experience also play important roles.

We will offer you a methodology for clarifying your expectations and determining the level of experience and skill necessary to be successful on the job; however, only you can envision the scope of responsibility afforded the individual as he or she performs critical tasks and activities. The level of autonomy you are comfortable with is often dependent not just on the job but also on how much you, the supervisor, will trust the employee as well as on your specific management style.

Previous managers can strongly color how individuals with the same level of education and experience will respond on future jobs. A controlling supervisor can sometimes limit the individual's ability to perform both now and in the future. Just as an elephant, used to a tether, will make no effort to roam free, individuals who have been severely constrained by controlling managers are ill-accustomed to working without supervision. They may be fully able to, but through their experience they have developed limits. These limits will differ from those whose experiences were shaped by less controlling, more free-wheeling managers.

Managers react to their subordinates in predictable ways. They also expect the employee to respond in a prescribed manner. The individual used to working with only limited freedom will seem to bother the free-wheeling supervisor with an endless array of "silly"questions and have a need to verify every little detail. This can annoy the supervisor. The supervisor may even interpret this careful attention to managerial direction as a sign of limited competence or lack of confidence. It is simply a display of employee conditioning.

On the other hand, a controlling manager will complain that an employee, used to working independently, goes off in his or her own direction, seems secretive, or is the proverbial loose cannon. The problem may not rest with the individual but rather represents a clash of styles rooted in job experiences. Both employees are simply responding in a manner that they have learned as appropriate. In long-term working arrangements, individuals either adapt to the manager's style or seek other employment. With the contract employee, working on a limited assignment, there is little, if any, time for this adaptation to take place. The fit must be immediate.

The manager, anticipating using contract staff, should identify his or her management style even before beginning the recruitment process. Since the successful contract engineer must mesh immediately with the supervisor's management style, the manager should consider up front how closely he or she manages and how independently the contract employee will work. The contract engineer is more than just an urgently needed set of skills. The engineer will also have a work style.

There is no "right" managerial style. Understanding your own style, however, requires that you take a mature look at yourself as a supervisor and at your work environment. Some organizations, such as those in the nuclear power, are anchored by regulation on careful managerial control of every step and process. Others, such as computer applications programming, are almost totally results focused and nurture freewheeling managers and empowered, independent employees. If you do not already know your management style, you should ask yourself: Do you run a tight ship? Do your employees report to you in detail and check with you as they move through their day? Do you guide or direct your employees? Do you provide them the tools to get the job done and then let them do it, or do (must) you closely monitor every step?

Your style and how you expect your employees to function must be communicated to the account executive and to any prospects submitted by the firm. A staffing firm that knows its employees well will already know the style and adaptability of the individuals it recommends. With some knowledge of your style, the firm can seek to submit prospects that will fit your style, or the firm may be able to provide coaching to help an employee adapt. You want the contract employee to come into your organization and become immediately productive, not spend a lot of time hassling with your way of managing. The firm wants a successful placement. You and the firm both have a stake in a successful placement, and you can win if you work together.

Looking beyond the Job Description or Summary

Most companies put their managers through a stringent justification process before they are allowed to add any new employees—whether contract or permanent. Depending on the organization, the manager must carefully examine the workload and current staff levels and capabilities, prove the need and select an existing (or develop a new) job description or summary for the new hire. Once this arduous process is completed, most hiring managers just assume that they have prepared the firm to recruit just the person they need. Unfortunately, this may not yield the desired results.

Frequently, the justification process is cut short. The manager, pressed for time, simply identifies a job title for the individual that matches the workload, selects a preexisting job description or summary that seems to fit, and seeks approval for the position before the workload—already running at near-flood stage—overruns the department. This is the reality of hiring, particularly when the new hire is going to be a contract employee.

The problem is that job descriptions or summaries do not tell the whole story. First, they are historical documents designed to help the organization fulfill a number of human resource needs. They are inputs and guides for compensation, performance evaluation systems, union contracts, career ladders, manpower planning, and succession planning systems. As components of these complex systems, they are developed within a consistent format and are deliberately broad and general in nature. In addition, the job description or summary is edited and reedited during a lengthy approval process, and as a result contains a lot of bureaucratic, legally and politically correct language.

One of the outcomes of the lengthy approval process is the unwillingness of many human resource departments to update job descriptions on a regular basis. Other concerns that keep job descriptions in use long after they should be retired include (1) fears that a change in one will trigger a chain reaction and lead to a wholesale revision of all the jobs in a "job family," (2) concern that new descriptions will proliferate and that jobs having similar duties may wind up with multiple titles, classifications, or FLSA status (exempt/nonexempt), and (3) suspicion that new descriptions are being requested to drive up the (compensation) evaluation of an existing job.

Given the reluctance to develop job descriptions or summaries as quickly as jobs change leads to a common condition. Almost without exception, they are out of date before they are approved. This is particularly true in fast-moving technical areas where new technology and improved processes are the order of the day. Therefore, job descriptions or summaries seldom reflect the dynamics of the work environment.

Also, because job descriptions or summaries serve many purposes (selection, compensation, etc.) and are expected to have "long lives," the language of these documents is intentionally general. For example, the description might say "produces engineering drawings." Or a modernized version might go as far as to include language like "develops CAD drawings." Is this information sufficient for recruiting purposes? Probably not. What the hiring manager really wants to know is, "Can the individual develop CAD drawings on our system?" In addition, the hiring manager may wish to know if the individual has had experience on a recent version of the software (e.g., AUTOCAD V.12, etc.) or if he or she has worked on 3D or solid models in addition to 2D drawings. The staffing firm and the prospective contract employee will not know that this is what is expected just from reading the summary. You must provide more explicit detail.

Another example of the inadequacy of job summaries and descriptions can be found in the area of nontechnical requirements. For example, every job description notes individual abilities in human interaction. You will need to look at what is on yours and determine if it really describes what is required. The summary might include language like "must be able to work independently." What does this mean in your company or even your department? Should the individual expect to be able to work for days completely alone at a computer terminal, or does it mean that the individual does not need help to find the cafeteria? Or the job description may say "works well with others." What does this actually mean in your work environment? If your company depends on teamwork, you will require individuals whose interpersonal skills are highly developed.

Third, the job summary is usually a laundry list of duties and responsibilities that fall within the scope of a specific title within a specific pay range. They often do not provide a shape for the specific job the contract engineer will perform. Although a duty or task may appear on the job summary, a contract employee may never in the course of his or her tenure with your organization actually be called upon to perform it. On the other hand, some other duties or tasks will consume the bulk of the individual's time. It is up to you to accent for the staffing firm those duties and tasks on which the contract engineer will focus.

The methodology that we recommend for building a job order so that you will get the contract employee you need is similar to an accepted methodology for job analysis, a standard human resource process. It includes clarifying the outcomes that you want as a result of bringing a contract employee on board (i.e., your objective), task definition, skill matching, interviews, and critical incident analysis. It can be used to develop new job descriptions or to select what is needed out of existing ones. In our approach, you are encouraged to focus on the specific attributes that you want in a contract employee if you are to have a reasonable expectation of having that individual meet your needs. If you have not yet justified the position, or it represents a new totally new position heretofore unused in your organization, you may consider using this methodology to develop not only the job order but a new job description that could fit a permanent or contract employee.

Whether you use the methodology to deconstruct existing summaries or descriptions or construct new ones, the desired outcomes of the work to be done and the list of tasks that must be performed for successful completion of the job or project should be your starting point for developing the job order for the contract staffing firm, *not* the preexisting job descriptions or summaries. Starting with a preexisting summary or description will influence you to use vague versus specific language and will probably lead you to include a number of job duties and skill and experience qualifications that are either unnecessary or inflated. As a consequence, you may find yourself answering a lot of questions from the staffing firm, interviewing a large pool of prospects who are not "on target," and having a difficult time filling the much-needed position.

You may ask, "When do I have time to do this analysis? I barely have time to put together the justification. I need the person now." This is one of those situations that merits an offensive directive cliché: "Make time." If you spend just a morning or an afternoon putting together the information needed for a first-class job order so that the staffing firm can then recruit and submit candidates who will rapidly meet your approval, you will have spent less time in the long run.

The difference between the shortcut and the approach requiring a modest time investment is not unlike shopping. If you need a new pair of shoes and have clear specifications (brown, lace-up, size 8), you can go into a shoe store and quickly narrow the selection and choose the shoes you want. In the hiring situation, the staffing firm will narrow its candidate choices to those who fit your description. This allows you to quickly check for fit and make the hiring decision.

The scenario is different if you simply ask the clerk in the shoe store for shoes. If the clerk has to check for your size and bring you everything in the store that fits for your review, you will be lucky to conclude your business before closing time. If you follow the same process when shopping for a contract engineer, you will interview numerous candidates and waste considerable time trying to find that magic person who fits.

Building the Job Order:
Objectives, Tasks, Skills,
and Experience

What follows are the specific techniques that you can use to construct a job order. A simplified case example can be found in the appendix, in which a hiring manger and a human resources professional develop a job order through a process that includes job description deconstruction.

Setting Clear Objectives

The technique recommended combines some of the traditional tools of job analysis in a form that will meet the needs of supplying contract engineering and technical staff. The first step is predicated on the assumption that you have a clear picture of what you expect the contract engineer's endproduct to look like. Therefore, you should translate your goals or project objectives into a basic job responsibilities.

For example, if you are hiring a contract technician to write the quality manual for your division, you know that the product will be a series of printed documents relating to quality control. Your next task is to develop a list of all of the job activities or tasks that you will expect this individual to perform in the pursuit of the stated deliverables.

Listing Specific Duties and Tasks

This part of the task is more difficult if you are developing a job order for a job that has never been done before in your organization. To compile this information, you may need to interview individuals working in your organization who have done similar work (e.g., documentation of any nature), or you may wish to call colleagues who work in other companies that have produced similar products. Once you have some idea of the tasks involved in getting to the desired result, you can turn to an existing job description or summary for guidance. The language, while general, may help you complete a list of all the tasks that must be performed to get the manuals written. Depending on how your job descriptions are formatted, you may actually be able to simply expand each of the areas with your list of activities.

Specifying Skills and Experience

Once you have identified the critical tasks and activities, you need to specify areas of expertise the individual must possess. Once again, it is recommended that you not turn immediately to the existing job descriptions or summaries since the skill and experience requirements are often not limited to the immediate job but may anticipate the development of the job holder to more advanced jobs in the same job family. In addition, the skill and experience requirements anticipate that the job holder will have to fulfill all the requirements set out in the job description or summary. If the tasks that are included on your job order represent only parts of what is included in the broad job description, the skill and experience requirements on the description probably exceed your needs.

In addition, the skill and experience requirements for the job order need to be specific enough to help the contract agency provide someone who can perform at the required level on day 1. Will the contract employee need specific knowledge of the manufacturing processes in your industry? Will he or she need to be familiar with the specific technology that you use? Will he or she need to know how to use special software?

When a contract engineer is hired for a specific project, you must be careful to identify specific tasks that must be accomplished by the individual throughout the engineer's tenure on the project in order to build a comprehensive list of necessary skills and experience. You may be able to generate the list simply by reviewing your project plan and noting where you have already identified unmet needs. However, it is all too easy to be swept up in the requirements and needs of the early stages of a project—particularly if it is running behind schedule—and not think about what lies beyond your immediate needs.

For example, knowing that the quality manuals must be written to meet a deadline, you bring on board an excellent quality technician, with experience in technical writing. Your company, however, expects to print these manuals out of house. Several weeks into the project you realize that your contract technician has no understanding of printing technology. This will create difficulties in completing the project. The contract specifications in the job order should have included not only technical writing but also "knowledge of and demonstrated experience with printing technology." Or, more specifically, the job order could have said "experience with desktop publishing and offset printing technology, including a working knowledge of the following software packages."

This is an extremely simplistic example, but you would be surprised at how often managers needing contract employees are blinded by their most immediate need and fail to specify skills needed later in the project. They then become dissatisfied with the individual as the work moves into areas that they did not carefully define before hiring.

Don't eliminate items from consideration until you have visualized the project from start to finish. The result will be a laundry list of what it will take for the contract employee to successfully complete the assignment. Your initial list may be as short as 8 to 10 items, or as long as 50 or 60! However, it is easier to pare items from a long list than to add items on the day that the job order has to go to the agency.

Special Considerations in Assigning Contract Employees to Teams

When you are hiring a contract engineer to be part of a team, you will need to consider the role the individual will fill within the team. For example, if you expect the contract engineer to work on a team charged with implementing a major manufacturing change that requires the installation of robots and special materials-handling systems, you will need to interpret the specific role in the team that the contract manufacturing engineer will play and the tasks performed. Members of cross-functional teams may not need every skill required for the completion of the project. You may be able to reduce the required tasks and the expertise needed by the contract engineer if he or she is working on such a team.

Refining the List: Tasks, Skills, and Experience

Once you have completed your list, you should go back and rank each item for its importance in meeting the project's needs (1 for those most important, 2 for those of moderate importance, 3 for those of least significance). As you rank the items, you should be sure to give a 1 to any that are unique to this position. This is particularly important if the contract engineer is to bring to the project specific expertise unavailable elsewhere. This process will clarify any special skills that you will need to include in your job order. You can also use it to give you a clearer sense of what the most important duties, tasks, and responsibilities that the engineer will perform as well as the required qualifications. From this, you can give the staffing firm and prospects a realistic preview of the job.

If you are writing a new job description, use the prioritized list to develop the general language of the summary. You will also want to begin developing performance measures that you will use to evaluate the contract employee's success on your project. If you were developing a job description for a permanent employee, this would be a key element. However, since the employee is a contract employee, performance evaluation will take another spin. Chapter 10 will detail the manager's role in evaluating contract employees.

If you are using an existing job description or summary to check for completeness in writing a job order, you can use the ranking process to identify the most important groups of tasks, duties, and responsibilities that you expect from the contract employee. However, because you began the process by listing your overall objectives for the position and the specific tasks of the target project or job, you will by this point be well beyond the generalities of the job summary and into a level of detail that will allow the staffing firm to secure the employee you need.

Clarifying Cultural Factors for the Job Order

There are other issues that you must consider before you are ready to finalize your job order. As you developed your key elements of success, we suggested that you might need a catch-all category—fit or culture. What initially appear to be tag-along elements can strongly impact the contract employee's success. Imperfect cultural fit can doom the success of a contract engineer. Here are some cultural issues that you will need to be prepared to address during the recruitment process.

Composition of Your Workforce

In many instances the workforce itself defines the organization's culture. A family-owned small business is often characterized by the personalities of the family members. These businesses are sometimes plagued by family rivalries and other dysfunctions. Even within very large companies individual units will have their own unique cultures. For example, a plant in the rural heart of the Midwest, staffed primarily with hometown talent, will reflect the values and interests of a small-town farming community. There will be a greater focus on family-driven activities.

If your organization heavily invests time and psychic energy in hometown football, the company softball team, or company picnics and golf outings, you will need to communicate this to your recruiter. A contract employee with a big-city attitude and an indifference to the activities and values of the area will find it more difficult to fit in. The ability to fit quickly into the culture of the organization is essential for the contract employee.

Although federal laws prohibit discrimination based on age, race, sex, religion, and national origin and although contract employers are prohibited from discriminating against employees of the staffing agencies on these bases, they cannot legislate the minds of the individuals in your department. You should have a realistic view of your employees' tolerance for diversity. If you are a manager responsible for a unit that needs remedial diversity training (the equivalent of social Neanderthals), you can expect potential friction if you bring in an engineer who does not fit the unit's profile. If the "boys" in your unit still refer to women as "gals" or "girls," the contract staffing firm should at least be warned before sending you a woman engineer. Similarly, a unit composed heavily of late-career permanent employees may have difficulty accepting a "young whipper-snapper."

Smoking often creates workplace conflict. If your company still allows smoking, you will want to let the staffing firm know this before it recruits you a militant nonsmoker. Similarly, if your group is dedicated to a particular health-driven work life—everybody goes to the gym at lunch—an overweight couch potato may be lonely and poorly accepted throughout the contract. These are the unfortunate realities of human interaction. In no way should they preclude your hiring someone who does not fit the profile; however, you, the staffing firm, and the engineer will be best served if there are fewer surprises.

Corporate Culture

Although workforce composition is an important component of the organization's culture, many organizations have taken great efforts to define how the corporation relates to its publics. If your firm has a written mission statement, examine what values it embodies. You will want employees, contract and permanent, to be in tune with them. You should realistically ask yourself if the company actually adheres to the mission statement or is it treated as a pretty piece of calligraphic art for the reception area.

Ask yourself whether your organization has a global or local focus. If you are globally focused, a contract engineer with experience in global, multinational corporations will fit more rapidly and understand the marketplace your business works in. Does your organization value those who question or those who follow? Depending on your organization's preference, you will not want a contract engineer who will question early and often or conversely one who just says yes.

What is the predominant approach to jobs and work? If your culture is one of entitlement, just bringing a contract employee on board may meet resistance. On the other hand, if your culture is one where the work is strongly driven by a passion for making something happen, you will want to recruit a contract employee with a passion for action. If your employees have a hard time knowing when the day ends because they are so enthralled with their projects, you will need to make your account executive aware of this before you recruit.

If you plan to recruit a contract engineer for a job with potential hazards, you should look at your company's approach to workplace safety history: Are you better than average? Is it a priority? You may find your people unwilling to work or accept someone they view as unsafe. Your staffing firm can help you recruit an engineer who will fit.

As you begin drawing a mental picture of how the contract engineer will fit with the overall work environment, you should also think about specific interactions. Since most contract engineers are hired for projects, you should identify who the engineer will link with on the project. Just as the elements of the strategic supply chain were an essential part of developing a strategic linkage in Chapter 8 (pgs. 159-163), so too are the individuals in a project. Each employee supplies information or other work into another area of the project. The strength of the project team and often the success of the project hinge on these links.

In developing a picture of your ideal contract engineer, take a moment and think about whom the engineer will receive information from and provide it to. You will want the contract engineer to interrelate especially easily with these two connecting points. They should view the work from the same perspective. Sometimes the differences in view will be subtle. For example, an electrical engineer working on developing a new power and motion control system will come at the engineering problem slightly differently than an engineer with a specialty in hydraulics. The electrical engineer will think in terms of weight, current, electric, and electromechanical solutions, while the mechanical engineer with a deep experience in hydraulics will look at pumps, lines, pressures, and flow. These two engineers must be able to speak to each other in terms that they both understand, and they must develop an appreciation for the experience and the point of view of their coworker. Similarly, a quality technician and a computer programmer may have to work toward a common language as the computerized quality-monitoring system is being developed.

The ability to communicate with others on the project comes from not only sharing an understanding of the other's perspective but also from some hard knowledge of the other's field. This can be obtained either through experience in your facility—the case for permanent employees—or through careful selection of contract employees. Once you have put together the specifics of the position, you should be ready to discuss it with your firm account executive.

Submitting the Job Order

Finding a staffing firm that can meet your needs and developing a clear picture of the contract engineer you need are the first two steps in getting the right person for the job. The last step is effectively communicating your needs to the staffing firm account executive. Most staffing firm account executives are trained to ask for specific information that will help them fulfill your job order. Here are the types of questions that you might be asked to field.

What are the duties, scope, and responsibilities of the job? It is very important that you communicate more than just the job title and the generalities. If you have analyzed the job as we recommended above, you will be well equipped to provide this information with specifics that will ensure that the firm understands the requirements of the job. As you detail the requirements of the position, expect to answer questions on whether the position involves travel. If so, be prepared to give a percentage. "Occasionally" may mean every 3 months to you or 3 days a month to someone else. Similarly, you should also mention if the individual will be subject to overtime. Be prepared to identify who will be the direct supervisor.

When, where, and how long will you need the staff? Before you contact the staffing firm, you should know how long the assignment at your firm will be. Although you may have a clear idea of when you are expected to complete a project, you should also make sure you know how long you will need contract staff. If you are uncertain, you should let the staffing firm know up front that the position might be subject to extension. Staffing firms with highly skilled, in-demand employees are keen on knowing in advance when one contract will end so that they can place their employees immediately on another contract. If there is a question about the time frame, you will need to keep close tabs with your account executive. Similarly, you will need to let the firm know if you are staffing for several locations or expect the individual to work at a different location from your own.

Why is the position open, and how long has it been open? The firm will need to know if the position is project-based or the result of turnover in your organization. The firm will want to know if your organization has been undergoing a restructuring or a reengineering that has impacted the position. If it is project-based, you will want to give the recruiter a sense of the size, extent, and duration of the project, even if you are not planning to use the contract person for the project's entirety.

The firm will also need to know if you have already made other efforts to fill the position. If you have already interviewed numerous unsatisfactory candidates or even used other firms' personnel, you should let the account executive know the situation. Although you may wonder at the wisdom of telling the staffing firm that you have recruited for 3 months to fill a vacant position and can't or that you have already contacted two other staffing firms to look for the right person, the staffing firm should know what it is going to be up against in finding you the right candidate.

It is particularly important to give the staffing firm some history on your search if you have already interviewed internally and externally and/or have a candidate in contention for the position. This can prevent any misunderstandings at a later date. The staffing firm will not want to submit candidates that you may already have interviewed, or if it places a contract person in the position, it will not want to have the history create difficulties with other staff.

What is the anticipated pay range? Most agencies will be interested not only in what you intend to pay the contract employee but also in what the range is for comparable employees in your organization. Since compensation is often a function of market demand and may vary considerably within the same geographic area, you must go beyond "We pay competitive wages." You may also want to provide information on benefits that your company provides its employees since they form part of the complete compensation package. Today, company-paid medical insurance is a benefit that has a real dollar value. If your company pays just marginally competitive wages but has an enhanced benefit package (paid relocation, company-paid medical insurance, etc.) that attracts employees, you should let the staffing firm know about these benefits.

What is the hiring process? Since interviewing prospective candidates is central to the process, you should give the staffing firm specific information on who will need to interview the candidates and when these individuals are available. Since some of your prospects may already be working, perhaps at some distance, you should also include whether you are free to interview in the evenings or the weekends. If you know in advance that you will be unavailable for specific dates (either on vacation or company business), be sure to tell the recruiter. This is particularly important if you are the key decision maker, need the person immediately, and anticipate a travel-filled calendar. You, the staffing firm, and the candidates will need to find mutually accept-able dates.

If more than one individual is going to take part in the process, you should let the staffing firm know how many interviews are required, where they will be conducted, and whether you will also need to test the candidates. If multiple individuals are going to participate in the inter-view process, you should also find tentative dates when they are avail-able to see candidates. You do not want to experience this all-too-famil-iar scenario: In this nightmare, you interview absolutely the best candidate for the position. The other interviewers are not available for an extended period of time during which the candidate finds another position.

You should also give the staffing firm a sense of how long it will take your organization to make a decision on a candidate. How many days or weeks will it be from the first interview until you make an offer? This gives the staffing firm valuable data for managing the candidates dur-ing the process. Time drags mercilessly for a candidate eagerly waiting to hear from a prospective employer. If the staffing firm and the candi-date know they can expect to hear from you in a specific time frame, the wait is easier. With a clear time line, you will find the staffing firm per-sonnel to be your ally, not someone you want to avoid as he or she checks up on where you are on the decision.

When is the person expected to begin work? It is important that you let the staffing firm know when you expect the contract employee to report for work. This is particularly important if there will be a significant lag between when you make the offer and when actual employment begins. If you are staffing a project with a remote start date, the staffing firm will need to know this before it begins recruiting for you. Also, if your situation is such that you will need the chosen candidate on site almost instantly, let the staffing firm know.

Throughout the recruitment and interview process, the key to success is to maintain an open and honest dialogue with the account executive. If you interview a candidate who does not meet your needs, be sure to give the account executive enough data on why you did not choose the individual so that he or she can refine the search based on mutual knowledge of how you reacted to a candidate. It is important to remem-ber that the staffing firm is eager to make the placement, and the only way this will occur is to provide you with candidates who will meet your approval.

Summary of Key Points in This Chapter

1. Successful recruitment of contract staff requires selecting a top-notch staffing firm and then communicating specific needs effectively.

2. Training and experience determine the level at which employees function. It is difficult but essential that employers clearly identify performance expectations for contract employees.

3. The scope of an employee's responsibility that he or she assumes depends on the individual's conditioning and the supervisor's management style. Ascertain your style before you recruit.

4. In developing the job order, begin with your objectives, the tasks that must be completed to achieve those objectives, the duties and responsibilities of the individual brought in to do the tasks, and the knowledge, skill, and experience that should enable an individual to successfully meet your expectations. Prioritize tasks and the related knowledge, skills, and experience to guide those who will be seeking candidates to fill your job order. Use your company's job descriptions or summaries to refine the language and to determine whether you have addressed all critical areas. Try to bypass generalities for specifics whenever possible.

6. Include culture and other determinants of fit including the workforce composition, corporate culture, and workplace communication linkages in the job order, or in other communications with the contract agency.

7. When submitting a job order to a staffing firm, be prepared to give details on the job, the expected compensation, and the interview and hiring process, as well as any efforts already made at recruitment.

10
Evaluating the Contingent Hire

Returning from lunch, Frank, an account executive with a contract staffing firm, really didn't know what to expect when his voice mail had an urgent message to contact Morris at Northeastern Plastics. He had recently placed two contract engineers with Morris's department. The engineers were highly skilled professionals with several years' contract experience. When he spoke with both engineers in recent weeks, they had expressed concern with the contract's progress.

When Frank returned the call, the news was not good. Morris seemed quite agitated. He kept saying, "I just don't know what I am going to do. These guys just aren't working out." Frank, who had previously successfully placed a number of staff with Northeastern Plastics, asked Morris what seemed to be the problem. He wanted to get to the root of Morris's dissatisfaction so that he could immediately rectify the situation. As he gently probed, all he could find out was that the work in the department was still way behind. The two contract engineers seemed to fit well and could do everything that was asked of them. Frank, still puzzled, asked for more details. Morris could not provide them, but he did know that what he wanted was "two guys who can get the job done." Before the call ended, Morris had agreed to keep the two engineers until Frank could get him some additional candidates to consider. Frank hung up the phone and pondered, "What is going on? Why can't Morris pinpoint what his problems are with the engineers? Is there a real problem?"

Although communication with your contract staffing firm is key to obtaining contract employees who will meet your needs, managing and evaluating performance during the contract requires ongoing honest and open communication. Performance evaluation was not even on the radar screen when contract employees were 2-week fill-ins replacing vacationing clerical workers. What passed for appraisal was usually done after the fact and reflected on the firm, not directly on the employee. The staffing firm was judged on its ability to provide an employee that could do the job. The temporary employees were virtually faceless commodities. If they couldn't do the job or didn't fit in, the staffing firm just sent a replacement. Satisfactory performance of temporary personnel resulted in additional hires for the staffing firm. As contracting has changed, so too has the need for individual performance appraisal. Today, contracts are longer and involve higher-skilled staff across a broader range of occupations. The contract employee today is expected to work with less direct supervision and has the potential for greater impact on the organization. In this chapter, we will consider the challenges of evaluating contract employees, the typical performance appraisal process, and the roles of the staffing firm and the contractor.

Performance appraisal can be visualized as three segments of a continuum. All managers using contract staff can expect to be involved at the minimum in a three-step performance appraisal process: (1) when the job is defined and performance standards are set, (2) during the performance of daily work, and (3) annually or at the conclusion of the contract.

The managerial demands of each segment are different, but they all link together to form an effective performance appraisal system. The first phase is, as in so many other managerial activities, a planning phase. At this time the manager sets expectations and standards as well as outlines and communicates the employees' goals. The second formative phase is by far the longest. During this phase the manager directs, guides and coaches, and provides the motivation that results in the employee's performance. Many managers overlook their responsibility toward employee development and appraisal as they struggle to get tasks completed and problems resolved. The third, summative phase is the review of the performance at a prescribed interval. This culminates in the annual review. It is important for the manager with contract staff to understand his or her role relative to these employees throughout each phase of the continuum. Managing the performance appraisal of contract employees is key if your organization works in a strategic alliance with a staffing contractor. Just as performance systems enable companies to build strong, resilient, productive workforces, a productive performance appraisal system for contract employees will strengthen the links between the contractor and the staffing firm. This will enable the staffing firm to provide employees who will meet performance expectations.

More Than Just an Annual Review

Every employee expects to receive performance evaluation. A manager provides performance appraisal data every time he or she approves a drawing, gives the go-ahead on a project plan, or provides guidance and direction during a progress meeting. The appraisal of employee performance is just as much a part of the management process as providing guidance, direction, and motivation. The manager who communicates performance data regularly to employees, offers coaching, and identifies and remediates skill deficiencies on an ongoing basis may find the much-vaunted annual performance review anticlimactic. These annual sessions become an opportunity for the manager and subordinate to mutually assess overall progress and set the stage for continued employee growth and organizational productivity. In the ideal, the annual evaluation should hold no surprises for either the employee or the supervisor. It should just be another part of a management continuum. Performance appraisal of the contract employee fits this model with the unique working relationships calling for some minor adaptation.

The contract employee's working relationship creates ambiguities and complexities in the performance evaluation process. As we have noted, the contract employee works on a triangular relationship with the staffing firm, the contractor, and the employee. The firm, as the contract employee's employer of record is responsible for the evaluation of its employees. The firm, however, does not have daily supervisory responsibility for its employees. The contract employee works under the direction of the contractor. The job site supervisor has far more direct contact with the contract employee than the staffing firm. The job site supervisor assigns tasks and evaluates the ongoing work. Since appraisal is more than just an annual event, both the staffing firm and the job site supervisor bear a responsibility for developing an accurate picture of the work performed and the employee's performance.

Because the firm does not have daily contact with its employees, the employer of record, the firm, must depend on information provided by the job site supervisor. The direct supervisor, therefore, has a significant role in the contract employee's performance appraisal and career development. Since the contract employee's economic value and subsequent compensation is based on skills, the firm needs to know about skills that the employee obtains and develops on the job. Although the employee may identify new skills for the staffing firm, the contractor can verify skills learned or document for the staffing firm deficiencies uncovered during the performance of the contract. Although staffing firms often provide ongoing training for their employees, they must rely heavily on the job site supervisor to provide the objective evaluation of the employee's ability to apply abstract learning on the job.

Since the actual on-the-job direction, work assignment, guidance, and approval are the responsibility of the direct supervisor, the staffing firm can react and work from performance data provided only by this individual. For example, unless the supervisor reports violations of work rules, tardiness, or absenteeism, the staffing firm cannot promptly mitigate the situation. Without information provided by the job site supervisor, the firm would not be aware if there were interpersonal conflicts between its contract employees and/or permanent employees of the contractor. It is impossible to stress too highly the need for the contractor to maintain a close link with the account executive at the staffing firm. This linkage is key to both high performance and effective performance evaluation.

Phase I: Setting the Expectations

Performance evaluation begins before the contract employee sets foot on the job site. It actually begins with the development of the job order. Dissatisfaction with contract employees is often a function of miscommunicated or inchoate expectations. In the last chapter, we outlined a process for developing an effective job order. The job order is key to satisfaction with the contract employee's performance. During the process, we suggested that you carefully consider the elements essential for your project's success. Use these elements to develop clear performance objectives around which you can judge the contract employee. If you already have performance measurements for the job, you should apply these to the contract employee. If you do not have these, you should develop a picture of what the successful completion of the job will mean and build performance objectives for the contract employee.

Since most contract employment is project-based, it is a straightforward process to frame performance objectives around the project's goals. Contract employment with its clear focus on a specific project lends itself to management by objectives. The project and by extension the contract feed into the organization's goals. The project should enable the department to meet either departmental-level goals or corporate goals; therefore, contract staff can be expected to meet objectives that lead to these goals. The manager must shape the contract employee's objectives as they relate to the project and its relationship to the organization's success.

We suggested in the previous chapter that you should clearly identify the "product" that you expect from the contract employee's tenure. The product may simply be an input into something else, but it is what you specifically envision the contract staff accomplishing. In the example in the last chapter, the quality technician was to create a series of manuals. Likewise, the manufacturing engineer's assignment was shaped around designing and implementing a new work cell. A network management engineer's assignment might include implementation of a corporate intranet. The manager hiring each of these contract employees can use these products to develop clear objectives for each employee.

An objective must be active, time-bound, and measurable. So an overarching objective for the quality technician's assignment might read "to develop (active) three (measurable) quality manuals in 6 months (time-bound)." There will be specific tasks that relate to this goal, and each will have its own similar objective. The accomplishment of the project and the satisfactory completion of its tasks should frame the basis for evaluation.

If you set the objectives before you begin the recruiting process, you will find that you have fewer problems recruiting individuals who can and will meet your expectations. Professional football teams at their recruiting combines use set minimum speeds and other performance standards to evaluate potential talent. Before drafting a rookie, the team has a clear picture of skill performance and recruit to meet their needs. Since you are employing the proverbial free agent via the contract staffing firm, you should try to recruit at a level that will meet specific performance expectations.

Even with a set of objectives and clear performance expectations, it is easy for the manager employing contract staff to fall into a series of performance evaluation traps. These unnecessarily will create dissatisfaction with the contract employee. We urge you to keep in mind the following: *You are hiring a skilled employee, not a genie.*

In a competitive environment that forces companies to run not just lean and mean but human resource anemic, many managers must wait until their department's workload is crushing before they can get the authorization for additional employees—contract or permanent. Consider for a moment the steps and the time that elapses from authorization, to recruit-ment, to scheduled interviews, on further to offer and acceptance and finally the employee's arrival. The steps remind one of the child's song about the old lady who swallowed the fly and then a spider to catch the fly, and all sorts of other creatures in succession. The song's refrain is "I don't know why she swallowed the fly." The manager can, like the old lady, become so caught up in the process of obtaining the employee that the workload that initiated the process can grow from burdensome to overwhelming. It continues to back up during the process with each day compounding the problem.

In these work situations, there is a brief moment of relief when the contract employee, Hope, arrives. Here is where the magical thinking, the source of dissatisfaction with the contract employee, begins. For some reason, the manager, like Morris in our opening vignette, and even the other overworked professionals in the department, assumes that somehow the contract employee(s) will make the backlog vanish instantly and restore normalcy to their troubled work situation.

All too often managers expect the contract employees to work like magic genies. "Poof!!" the defective logic goes, " Now that we have the help we need, the backlog will disappear." Even if you hired a genius, the contract employee is not a genie. Each contract employee can be expected to accomplish the work of only one individual. It is human nature to hope for quick-fix solutions; otherwise, why are magazine pages still filled with get-rich-quick schemes and miracle diet formulas? In situations in which a backlog has developed, set up a realistic timetable for whittling it away before the contract employee arrives on site. Revisit the calculation to ascertain if the backlog has grown during the recruitment process, particularly if it is protracted. Also, check to see if the situation that created the backlog—a new product introduction, a manufacturing change, or other situation—has abated or has been fur-ther aggravated. Ask yourself, "Is this as bad as it will get?" The situa-tion may be temporary, but work is organic and grows and wanes. Your estimates on your backlog, its growth and eventual resolution, should be part of your communication with the staffing firm and the contract employee during the recruitment process.

Second, the situation is different if the contract employee is being hired to bring a specialized talent to a new project team or into an exist-ing department where there is not a backlog situation. The employee's performance cannot be measured against a volume of work but rather must be framed against the performance measures outlined in the job order and or job summary. In these instances it may take time for results to show. Again, don't expect magic.

Reserve judgment until the person has had time to create the expected impacts. Also, particularly in situations that require interfacing with other areas, it may take some time for the necessary linkages to forge. For example, the quality technician may not have the information needed to draft the manuals for several weeks or even months. Even though the project plan might indicate specific dates for completion of each phase from data gathering to printing, the manager is tempted to constantly ask like a child on a car trip, "Are you there yet?" Unfamiliarity with the employee and his or her capabilities can create anxiety about the person's ability to deliver the desired product. Each time the contract employee says "Not yet!" it waters the manager's seeds of dissatisfaction. The anxiety increases and the manager's esteem of the contract employee dwindles. Avoid magic thinking by setting realistic project milestones and then stick with them.

Phase II: Evaluating the Progress

During the course of the contract, the job site supervisor will have numerous opportunities to evaluate the employee's progress and provide corrective advice. It is during this period that managers gather the information and provide the managerial direction and guidance that feed traditional performance evaluation systems. The annual performance review culminates and summarizes this process. For the contract employee, this phase of the continuum is either yearlong or lasts the duration of the project. Although Chapter 6 provided tips for managing the contract employee, here are some additional tips that relate directly to performance management and measurement.

Apply a single standard fairly to all employees. A new employee, whether contract or permanent, will need time to acclimate to the work environment. Contract employees are accustomed to adapting rapidly to new work situations and usually adapt quickly. Managers, knowing that they have hired the contract employee for a predetermined assignment, sometimes develop unreasonable expectations for how quickly the employee can come in and adapt. They expect instant results. This is unfair and can lead to undeserved dissatisfaction with the employee.

In Chapter 6 (pgs. 127–140), we focused on integrating the new hire. Before finding fault with the contract employee's fit with your organization, consider how much orientation you have given the employee. Have you outlined the parameters of high performance? Have you given the employee the tools for success?

Contract employees sometimes complain that their supervisors lurk waiting for their mistakes. They claim that they are then resoundingly criticized for mistakes or omissions that would be overlooked if committed by permanent employees. When evaluating a contract employee's performance, avoid using a double standard. This is also why it is so important to communicate your performance expectations to the firm and the prospective contract employee during the recruitment process. If you are using an existing job description as the basis for your job order, not only set but also apply the same performance standards to the contract employee as you might a permanent employee. This will help you avoid this trap.

Don't expect a contract employee to have skills you did not specify in your job order. In Chapter 9 (pgs. 182-183), we stressed the need to develop a job order that reflects all of the skills you will need for the duration of the contract. Since you are hiring skills in a contract employee, it behooves you to develop a complete inventory of the skills you will need and the desired level of competency for each. It is not the contract employee or the firm's responsibility to read your mind or interpret the needs of your project and produce what you did not order. Yes, sometimes you may get lucky and get more skills and even higher levels of competency than you anticipate. However, you cannot depend on this.

Your account executive will probe your needs and try to deliver an employee who will meet your specifications. You also have a responsibility to explore each provided prospect's capabilities during the interview process. Once the individual is on site and working on the contract, you should not expect to hold the employee accountable for your mistakes in developing the specifications. Frequently, when a manager tells an account executive that a contract employee can't do the job, the reason is that the manager did not specify clearly either the skill or the level of competency. Sometimes a skill mentioned as only peripherally required in the job order will actually be the contract employee's primary focus on the job. The staffing firm will have recruited a prospect that meets the specification—a candidate with only peripheral knowledge. Once on the job, the person's performance will not meet the needs of the contractor. The contract staffing business is highly competitive, and every staffing firm will try mightily to deliver exactly the employee that you specify.

Sometimes the labor market will make it very difficult to find prospects, even through multiple staffing firms, with a complete inventory of desired skills. These situations will test your managerial skills and creativity. When you developed the job order, you placed priorities on the skills and experiences most critical to the success of the project. When an individual cannot fill all of the skill package needed, consider the possibility that you may have to divide up the work, assign portions to other individuals, or even hire multiple contract employees for different phases of the project.

This is particularly the case when a lengthy project evolves and requires significant shifts in assignment. Managers develop a comfort zone with long-term contract employees and sometimes expect them to respond like permanent employees. A permanent employee will take a developmental assignment; however, you cannot expect this from a contract employee. Some staffing firms, like ITS Technologies, give their employees the right to request a change in assignment when the there is a shift in the contract specifications. Managers eager to continue working with a familiar contract employee become distressed when he or she cannot or does not choose to shift into a new assignment not specified in the original contract.

A contract employee, satisfied with the working relationship with both the contractor and the staffing firm, will look for opportunities to continue a contract. The contract employees know well that their performance will be judged on their ability to competently perform across a specific spectrum of skills. Their careers depend on this, and stretch assignments are filled with risk. If you are able to reduce the downside risk, some long-term employees will accept such assignments. However, you should not have the same expectations from contract employees as permanent.

Instruct, don't assume. In recent years, many managers have focused on reducing their head counts. The resulting workforce has not only longevity but has experienced little influx of new employees. As a result, some mangers have lost sight of what it takes to bring new employees into their department. Although Chapter 6 (pgs. 131-135) presented a number of management strategies for integrating the new contract hire, integration ties to performance evaluation. If you do not integrate the employee thoroughly into the flow of the department, he or she will not perform at the desired level.

It is particularly important to provide the contract employee guidance as to your management style and departmental policies and procedures. If you have a particular reporting procedure or other requirements, be sure to communicate them. The contract employee is not clairvoyant. For example, if you hold a weekly meeting and expect all your subordinates to provide progress reports in a specific format, don't be surprised or disgusted if the contract employee doesn't understand your procedure when you have not explained it.

Similarly, the contemporary office is the proverbial technological minefield. Every office has different procedures for handling the computer systems, fax, phones, and so on. It is unfair to expect the contract employee to simply "catch on." Before you call the staffing firm complaining about how poorly the contract employee is fitting in, consider what efforts you have taken to ensure the fit. You should instruct, not assume familiarity.

Provide immediate feedback. As the job site supervisor, you will be responsible for the employee's daily supervision and work direction. You should handle this as you might for a permanent employee. Contract employees are eager to succeed and expect to receive your direction. It is sound management principle to provide employees immediate positive and negative performance feedback. This information gives the employee an opportunity for corrective action. The contract employee is no different.

The situation shifts just a bit when it comes to handling extraordinary performance on both ends of the spectrum. This is when it is important to remember that the contract employee is not your employee but rather the staffing firm's. Since the staffing firm is the employer of record, you will want to protect this relationship. It is the staffing firm's responsibility to provide the employee's annual evaluation. You, as the job site supervisor, should document instances of extraordinary performance so that the staffing firm will know of the employee's success. The employee will appreciate this. Human performance research has documented the positive effects of praise.

If you find yourself wanting to publicly praise the employee for a job well done, consider writing a brief memo documenting this and submitting it to the staffing firm. Do it immediately. Do not expect to remember anecdotal incidents that occurred months in the past when the staffing firm asks you to participate in the employee's evaluation. It is human nature to remember only the most recent past. Most work situations are based on the here and now—What did you do for me today?

Managers are accustomed to writing up negative performance. When you are working with contract employees, you should document incidents carefully. For example, if your company has a strict policy of no smoking within 10 feet of the building and a contract employee persists in bending the rule by lighting up on the way out the door, your first step will be to advise the employee of this company policy. Second, you should observe whether it is being followed by your other employees.

Many companies have policies on their books but choose to selectively enforce them. You should avoid subjecting the contract employee to selective enforcement. However, if your permanent employees pay attention to the policy, document how you handle the situation—tell the employee a second time, and so on. Be sure to keep track of dates and times and specific pertinent information. Then, finally, if you report a contract employee's repeated inappropriate behavior to the staffing firm with your request for the employee's removal, you can provide the staffing firm hard performance data to use in their evaluation. Morris, in our opening vignette, had no hard data points to explain his negative perception of his two contract engineers. The staffing firm will willingly comply with your request for another candidate, but you serve them better by helping the firm develop performance information on their employees.

Give the staffing firm coaching data on remediation needs for their employees. Perhaps the employee's negative performance does not sufficiently offset the benefits you are deriving to merit his or her removal. In these situations, the staffing firm will provide coaching for its employees. You, however, should provide information that supports requests for coaching. If you intend to maintain a good working relationship with the employee, the employee should be aware of your request. For example, if you have identified a performance discrepancy related to skills that can be readily remediated by a rapid review or self-paced learning, transmit your concerns to both the employee and the staffing firm. The staffing firm, as the employer of record, is responsible for the employee's development.

Phase III: The Final Evaluation

At the end of each employee's contract, you can anticipate receiving an evaluation form. At ITS Technologies, each contract employee completes a self-evaluation. The account executive then sits down with the job site supervisor and requests input. This is compared to the contract employee's feedback. The account executive sits down with the employee and reviews and compares both evaluations. When an employee remains on a contract for more than a year, the individual is evaluated on an annual basis.

This portion of the performance evaluation process is significant for the employee's growth and development. Since the employee's compensation is tied to skills and experience, this process provides valuable documentation supporting performance in preexisting skills and the development of new skills or transferable experiences. The contractor, as the employee's direct supervisor, has an important role in this process.

Each party benefits from the completion of the performance appraisal process. The contract employee obtains documentation essential for economic growth, and the staffing firm obtains data not only about the employee's success but also about how you evaluated its offering. If you found the employee lacking, this will provide information about employee development needs. From the evaluation process, the staffing firm can develop a profile of the characteristics of a high-performance employee. Your evaluation gives the staffing firm a standard against which to judge how you will react to other contract employees you might receive from the staffing firm.

An objective and fair performance appraisal is the last managerial responsibility. The process is made up of many components: determination of the need for a contract employee, the selection of a contract staffing firm, the development of the job order, and the recruitment and management of the contract employee. For most effective use of contract employees, you should not neglect any of these steps.

Summary of Key Points in This Chapter

1. Visualize performance appraisal as three segments of a continuum: (1) the planning stage during which the job is defined and performance objectives set, (2) the daily supervision of the contract employee, and (3) the summative evaluation after the completion of the contract.

2. Evaluation is an ongoing process performed on a daily, not annual, basis as part of the supervisory process.

3. The contractor is responsible for the daily supervision and by extension evaluation of the staffing firm's employees on the job site.

4. Contractors need to communicate both positive and negative performance to the staffing firm.

5. Set clear performance standards before recruiting the contract employee. Then, communicate them clearly to all parties.

6. Set realistic expectations. The contract employee is not a magician.

7. Don't expect a contract employee to have a specific skill unless it was requested in the job order.

8. Appropriate orientation is key to high performance of contract employees.

9. Document an employee's positive and negative performance.

10. Be sure to complete the staffing firm's requested appraisal documentation promptly. Doing so benefits the employee, the staffing firm, and the contractor.

Resources

_____. *National Technical Services Association: Membership Handbook.* Alexandria, Va.: National Technical Services Association, 1994.

_____. "Human Capital: The Decline of America's Workforce," *Business Week,* September 19, 1988.

Barner, Robert. "The New Career Strategist: Career Management for the Year 2000 and Beyond," *The Futurist,* September 1, 1994.

Beck, Nuala. *Shifting Gears: Thriving in the New Economy.* Toronto: Harper Collins, 1992.

Belous, Richard S. *The Contingent Economy: The Growth of the Temporary, Part-Time and Subcontracted Workforce.* Washington, D.C.: National Planning Association, 1989.

_____. "How Human Resource Systems Adjust to the Shift Toward Contingent Workers," *Monthly Labor Review,* March 1989.

Bennis, W. G., and P. Slater. *The Temporary Society.* New York: HarperCollins, 1968.

Block, Peter. *The Empowered Manager: Positive Political Skills at Work.* San Francisco: Jossey-Bass, 1987.

Boyett, Joseph H., and Henry P. Conn. *Workplace 2000: The Revolution Reshaping American Business.* New York: Plume, 1992.

Bridges, William. *JobShift: How to Prosper in an Economy Without Jobs.* Reading, Mass.: Addison-Wesley, 1994.

_____. "The End of the Job," *Fortune,* September 19, 1994.

Brown, Donna. "Outsourcing: How Corporations Take Their Business Elsewhere," *Management Review,* February 1992.

Byrne, John A., with Richard Brandt and Otis Port. "The Virtual Corporation," *Newsweek,* February 8, 1993.

Castro, Janice. "Disposable Workers," *Time,* March 29, 1993.

Caudron, Shari. "Contingent Work Force Spurs HR Planning," *Personnel Journal,* July 1994.

Church, George J. "Jobs in an Age of Insecurity," *Time,* November 22, 1993.

Culotta, E. "Teamwork Is Key to Solving Complex Research Problems," *Scientist,* March 8, 1993.

Davidow, William H., and Michael S. Malone. *The Virtual Corporation.* New York: HarperCollins, 1992.

Dekleva, Sasa M. "CFOs, CIOs and Outsourcing," *Computerworld,* May 16, 1994.

Dennis, Helen, and Helen Axel. *Encouraging Employee Self-Management in Financial and Career Planning.* Conference Board Report No. 976. New York: The Conference Board, 1991.

Dent, Harry S., Jr. *Job Shock: How You Can Prosper during the Work Revolution Ahead.* New York: St. Martin's Press, 1995.

Doyle, M. F. "Cross-Functional Implementation Teams," *Purchasing,* March 10, 1988.

Doyle, M., and D. Strauss. *How to Make Meetings Work.* Chicago: Playboy Press, 1976.

Dyer, W. G. *Team Building: Issues and Alternatives.* Reading, Mass.: Addison-Wesley, 1977.

Ekholm, E. L. "The Ins and Outs of Consulting as a Career," *Chemical Engineering Progress,* May 1996.

Ettore, Barbara. "The Contingency Workforce Moves Mainstream," Management Review, February 1994.

Feldman, Daniel C., Helen I. Doerpinghaus, and William H. Turnley. "Managing Temporary Workers: A Permanent HRM Challenge," *Organizational Dynamics,* Autumn 1994.

Fernberg, Patricia M. "The Skills Shortage: Who Can Fill These Shoes?" (An Interview with Samuel R. Sacco, Executive Vice President, National Association of Temporary and Staffing Services), *Managing Office Technology,* June 1995.

Fierman, Jacklyn. "The Contingency Workforce," *Fortune,* January 24, 1994.

_____. "What Happened to the Jobs?" *Fortune,* July 12, 1993.

Gordon, J. "Work Teams: How Far Have They Come?" *Training,* October 1992.

Grace, Tim. "Fortune 1000 Firms Embrace Outsourcing: Integrators Getting Positioned for Sales Bonanza," *Computer Reseller News,* July 11, 1994.

Greenberg, Eric. "Upswing in Downsizings to Continue," *Management Review,* February 1993.

Halvey, John. "Strategic Outsourcing: No Longer a Last Resort," *Information Week,* August 1, 1994.

Halverson, C. B. "Managing Differences on Multicultural Teams," *Cultural Diversity at Work,* May 1992.

Hammonds, Keith H., Keith Kelly, and Karen Thurston and others. "Rethinking Work," Special Report, *Business Week,* October 17, 1994.

Handy, Charles. *The Age of Unreason.* Cambridge: Harvard Business School Press, 1992.

_____. *The Future of Work: A Guide to a Changing Society.* Oxford: Basil Blackwell, Ltd., 1984.

Haneborg, Linda C. "A Productive Strategy for Managing Temporary Workers," *Benefits and Compensation Solutions,* InterMedia Solutions, July–August 1996.

Harkins, Philip J., Stephen M. Brown, and Russell Sullivan. *Outsourcing and Human Resources: Trends Models, and Guidelines.* Lexington, Mass.: LER Press, 1996.

Henkoff, Ronald. "Where Will the Jobs Come From?" *Fortune,* October 19, 1992.

_____. "Winning the New Career Game," *Fortune*, July 12, 1993.

Jacobs, Richard A. "The Invisible Workforce: How to Align Contract and Temporary Workers with Core Organizational Goals," *National Productivity Review*, Spring 1994.

Jamieson, David, and Julie O'Mara. *Managing Workforce 2000: Gaining the Diversity Advantage*. San Francisco: 1991.

Johnston, William B., and Arnold E. Packer. *Workforce 2000: Work and Workers for the 21st Century*. Indianapolis, Ind.: The Hudson Institute, 1987.

Kanin-Lovers, J. "Motivating the New Work Force," *Journal of Compensation and Benefits*, September–October 1990.

Kanter, Rosabeth Moss. "Collaborative Advantage: The Art of Alliances," *Harvard Business Review*, July–August 1994.

Katzenbach, J. R., and D. K. Smith. *The Wisdom of Teams*. Boston: Harvard Business School Press, 1993.

Kennedy, Joyce L. *Hook Up, Get Hired! The Internet Job Search Revolution*. New York: Wiley, 1996.

Kertesz, Louise. "Employee Leasing Firms Can Cut Some Benefit Costs; But Bad Apples Mean Buyer Beware," *Business Insurance*, March 29, 1993.

Kiechel, Walter. "How We Will Work in the Year 2000," *Fortune*, May 17, 1993.

Kilborn, Peter T. "New Jobs Lack the Old Security in Time of 'Disposable Workers,'" *The New York Times*, March 15, 1993.

Larson, Jan. "Temps are Here to Stay," *American Demographics*, February 1996.

Lenz, Edward A. *Co-Employment: Employer Liability Issues in Staffing Service Arrangements*, 2d ed. National Association of Temporary and Staffing Services, Arlington, Va, 1994.

McAuliffe, Claire. "Contingent Workers Deliver a Certain Advantage," *San Francisco Business Times*, June 11, 1993.

McKie, W. Gilmore, and Laurence Lipsett. *The Contingent Worker: A Human Resources Perspective*. Society of Human Resource Management Foundation, Arlington, Va, 1995.

Mason, Julie Cohen. "Workplace 2000: The Death of 9 to 5?" *Management Review*, January 1993.

Melcher, Richard A. "Who Says You Can't Find Good Help?" *Business Week*, January 8, 1996.

Moravec, Milan, and Robert Tucker. "Job Descriptions for the 21st Century," *Personnel Journal*, June 1992.

Morrow, Lance. "The Temping of America," *Time*, March 29, 1993.

Newton, Lucy A. "Stiff Competition for Talented Temps," *HR Magazine*, May 1996.

Nollen, Stanley, and Helen Axel. *Managing Contingent Workers: How to Reap the Benefits and Reduce the Risks*. New York: Amacom, 1995

Nye, David. *Alternative Staffing Strategies*. Washington, D.C.: Bureau of National Affairs, 1988.

_____. "Special Report: Rethinking Work," *Business Week*, October 17, 1994.

Parker, Glenn M. *Cross-Functional Teams: Working with Allies, Enemies, and Other Strangers*. San Francisco: Jossey-Bass, 1994.

Polivka, Anne E., and Thomas Nardone. "On the Definition of Contingent Work," *Monthly Labor Review*, December 1989.

Radeloff, Roger L. "The Need for the As-Needed Work Force," *Engineering Digest*, February 1995.

————. "Contract Workers: A Future Quality Management Challenge," *Continuous Journey*, June–July 1994.

————. "Contract Engineers: An Optimal Staffing Solution for Product Design Groups," *EMA Journal*, Winter 1995.

————. "Working for a Staff Contractor," *Chemical Engineering*, July 1995.

————. "Is Contract Engineering Right for You?" *Careers and the Engineer*, Fall 1994.

————. "The Supervisory Challenge of the Contemporary Employee," *Supervision*, March 1995.

————. "Have You Considered Contract Engineering?" *Chemical Engineering Progress*, May 1996.

Rogers, Beth. "Temporary Help Industry Evolving as it Grows," *HR News*, January 1995.

Samuelson, Robert J. "Are Workers Disposable?" *Newsweek*, February 12, 1996.

Savage, Charles M. *5th Generation Management: Integrating Enterprises through Human Networking*. Bedford, Mass.: Digital Press, 1990.

Stewart, Thomas A. "The Search for the Organization of Tomorrow," *Fortune*, May 18, 1992.

Tilly, Chris. "Continuing Growth of Part-Time Employment," *Monthly Labor Review*, March 1991.

Tucker, Robert, and Milan Moravec. "Do-It-Yourself Career Development," *Training*, February 1992.

Underwood, Anne, John McCormick, and Deborah Branscombe. "The Hit Men," *Newsweek*, February 26, 1996.

Waterman, Robert H., Jr., Judith A. Waterman, and Betsy A. Collard. "Toward a Career-Resilient Workforce," *Harvard Business Review*, July–August 1994.

Resource Guide

This *resource guide* has been prepared to provide the engineering professional seeking contract employment and the manager considering
employing contract technical staff an up-to-date source for staffing
information. There is a wealth of available information, in both printed
and electronic formats. This guide lists the Internet addresses of major
contract staffing firms and other useful career information sources.

Because of the Internet's broad availability and immediacy, this guide
focuses on Internet resources as opposed to those in the traditional volumes and directories available in many libraries. The development of
the Internet is rapidly changing how employers and prospects find one
another. With the Internet, both the job applicant and the hiring company can connect with one another in moments. The resources this
guide presents are at the reader's fingertips.

However, as those who have used the Internet extensively already
know, it can be time-consuming and often unproductive to spend hours
searching the Internet's vast resources for appropriate sites. The problem is compounded when the user must wait what feels like a very long
time for a selected site to transfer. Even more frustrating is to bring in a
home page dated 6 months prior only to find that it is little more than
an address and a short paragraph indicating that the site is under construction. To offset this frustration, we have chosen to provide a brief
description of the employer (as given in the site) and a review of what
the reader might expect to find in the site.

This list is also not all encompassing. It provides the reader the opportunity to access a selected group of resources rapidly with some knowledge of what each will include. Readers should note that the Internet is
a dynamic environment and that sites are added, changed, improved,
and even removed continuously. The reader may discover new
resources added since the completion of this guide in autumn 1996.

When listing contract staffing firms, we have chosen to provide the
reader the URL for the corporate "home page." Many franchise and
branch offices of larger firms have their own home pages. Most of these

are available from the corporate locator guides. To reduce confusion, we have chosen not to list each branch's URL. These are also most likely to change as offices relocate, close, and undergo normal business evolutions.

We have also chosen to include several large Internet resources dedicated to employment and career searches. These may include permanent as well as contract listings. It should be noted that although we have chosen to provide these listings, in no way does this constitute an endorsement or recommendation of the firm or site represented.

Staffing or Service Firms Employing or Providing Technical Personnel

AccuStaff: *http://www.accustaff.com*

Description.   On November 15, 1996, Career Horizons, headquartered in Woodbury, New York, is a national organization of 577 company-owned, franchised, and associated offices merged with AccuStaff of Jacksonville, Florida. From offices in 45 states and the District of Columbia, this firm serves mary markets. It is traded on the NYSE.

Features of the Site.   A very large site that provides a wealth of information on the firm, its divisions, and its services. It gives information for investors, potential franchisees, and prospective employees. There is an office locator by service type. It provides press releases and other industry information. It includes information on how to submit job orders and résumés online.

Adecco: *http://www.adia.com*

Description.   Formed in September 1996 by a merger of Swiss-based Adia SA and the French company, Ecco SA, Adecco is one of the world's largest suppliers of temporary and full-time personnel. With more than 2,500 offices in 40 countries and with 10,000 full-time employees, the firm employs nearly 1,500,000 temporaries yearly worldwide.

Features of the Site.   Provides information on the company, its history, and its organization. It gives news of the corporation and a listing of the jobs available. It includes an extensive and useful list of links and Internet sites that deal with human resources and job seeking. This section includes links to many job search sources online such as *Career Magazine*, career resources, and employment directories and databases, U.S. and European. It provides links to résumé writing tips and gateways for human resource professionals. The site gives extensive directions for locating information on compensation and government regulations on human resource matters.

Aerotek: *http://www.aerotek.com*

Description.   Founded in 1983, it now has offices in more than 100 cities, and serves hundreds of Fortune 1,000 corporations throughout the United States, Canada, and United Kingdom. Originally focused on staffing for the aerospace and defense industries, it has expanded its scope and now offers staffing services for firms in manufacturing, telecommunications, automotive, data services, energy and environment, information systems, and medical science.

Features of the Site.   Provides company description, worldwide location guide, divisions of Aerotex, client benefits, "what we can do for you," and contact information.

Butler International: *http://www.butlerintl.com*

Description.   With a corporate history dating back to the 1940s aircraft and defense industries, it now serves more than 1,400 customers in over 25 countries, has 50 offices worldwide, and annual sales in excess of $400 million. Today, it serves a broad industry mix (telecommunications, electronics, electronics, aerospace, and energy) offering technical personnel in over 1,000 classifications including designers, engineers, drafters, system analysts, programmers, technicians, and craftspeople. It provides technical personnel, project management, and strategic outsourcing services.

Features of the Site.   Provides company description, including some information on methodology for how it helps clients rightsize, press releases on the company, and office locator guide. Offers readers the opportunity to view current openings and to submit résumés and client job requests.

CDI Corporation:
http://www.cdicorp.com

Description. Based in Philadelphia, Pennsylvania, CDI Corporation has more than 40 years' experience in staffing and consulting for engineering, design, scientific, and technical needs. It has 250 offices worldwide and 150 in North America, and 28,000 employees worldwide. It is publicly traded on the New York Stock Exchange, and it has gross annual revenues of over $1 billion. Through CDI Information Services, it supports data processing needs from contract personnel to outsourcing.

Features of the Site. Includes description of the firm, what's new as well as listings of jobs, office locator, and contact information. It has résumé submission capabilities.

CTG (Computer Task Group):
http://www.ctg.com

Description. This 30-year-old firm was founded in Buffalo, New York, by two former IBM employees. Publicly traded since 1969, the firm has remained heavily committed to a range of computer services including platform migration and industrial systems integration. It has now penetrated 80 percent of the top U.S. computer markets. Through Profession Software Services (PSS), it fulfills contracts for professional staffing. CTG provides services through 55 offices in North America and Europe. The firm currently employs over 5,000 computer professionals.

Features of the Site. An extensive site that gives a history of the firm and investor information as well as providing the job seeker links to office locations. Each office location indicates the specific competencies sought. There is information on contract employment and variable staffing in information technology. There is an online newsletter, *The Technologies Digest,* and lists of sponsored job fairs and open houses and benefits information. It has an online résumé submission capability.

CTS Technical Services, Inc.:
http://www.ctstech.com

Description. Established in 1984 in Europe, the firm provides highly skilled technical personnel to the software, telecommunications and aircraft industries. In 1985 it established a U.S. operation in New York; it subsequently moved to Seattle in 1991 to provide service to the West and Pacific Rim.

Features of the Site.   Includes corporate history, a listing of available jobs, benefits of working for and with CTS Technical Services, Inc., overseas opportunities, and a list of clients. Gives résumé submission information but does not include an online résumé submission form.

Computemp:
http://www.computemp. com

Description.   Established in 1984 and based in Boca Raton, Florida, it is dedicated to providing staffing for the computer industry. It has more than 20 offices across the United States.

Features of the Site.   Includes current corporate information, office locator, local office news, tips and advice on résumé development and interviewing techniques, information for applicants and for businesses, franchising opportunities with Computemp, and an online listing of job postings. It has an online job application capability.

Computer Horizons Corp.:
http://www.chccorp.com/

Description.   Headquartered in Mountain Lakes, New Jersey, this firm operates 35 offices nationwide and employs 2,500 people, 2,200 in information systems. The company focuses its services on Fortune 500 customers and government contracts and currently services over 400 clients. One-third of the professionals with this firm have been with it for over 10 years. Publicly held since 1972, it is traded on the NASDAQ. Revenues in 1995 were $200 million.

Features of the Site.   Provides information on the firm, its history, service offerings, and management and financial performance. It provides phone, fax, and e-mail information for contacting its 35 offices. It does not provide résumé submission instructions.

DLD Technical Services, Inc.: *http://www.dldtech.com*

Description. Based in Cleveland, Ohio, this firm provides engineering, design, drafting, and computer programming services to firms across the United States. It provides staff on a weekly or project basis.

Features of the Site. Provides basic information on the firm and its services. It gives information on employment opportunities, contract and permanent, and an e-mail address for submission of résumés. There is no online résumé capability.

Dunhill Personnel System, Inc.: *http://dunhillstaff.com*

Description. Based in Woodbury, New York, with over 120 franchised and company-owned offices throughout the United States, Canada, the Caribbean, and Hawaii, Dunhill is one of the largest staffing providers. Since 1952, it has placed more than 2 million people in career opportunities. Dunhill provides search, placement, contract, and temporary services. Handles financial and administrative, health care, information technology, sales and marketing, and technical, manufacturing, and engineering personnel.

Features of the Site. A large site that gives information on Dunhill's office and franchises. It has an office locator guide with e-mail addresses and information on preparing for an interview and on how Dunhill recruits for permanent and temporary positions. It lists services provided and contact information by region and recruiting specialty.

Engineering Corporation of America: *http://www.ecofa.com*

Description. Established in 1966 and located in Seattle, Washington, this firm provides contract engineering and technical personnel in the Northwest and nationwide. It provides a range of personnel in software, mechanical, civil, manufacturing, environmental, hardware, electrical, structural, quality-reliability, and electronics disciplines.

Features of this Site. Provides information on this firm, a list of job openings in the Great Northwest, the benefits this firm offers contractors, electronic résumé submission capability, a quick information form for submitting skills, e-mail time card system for current employees, and hot links to other areas of interest.

IMI Systems, Inc.:
http://www.imisys.com

Description. Founded in 1979, this wholly owned subsidiary of Olsten Corporation is a $150 million provider of information technology services. Based in New York with over 1,500 specialists, this firm serves the manufacturing, banking, brokerage, insurance, computer, and telecommunications industries.

Features of the Site. Gives information on the firm and the services it provides. For the job seeker, it gives a list of available opportunities and has online résumé submission capabilities.

ITS Technologies, Inc.:
http://www.itstechstaff.com

Description. ITS Technologies, of which the author Roger Radeloff is the president, is a technical staffing provider based in Toledo, Ohio. With offices in Ohio and Michigan, this firm provides professional and technical staffing and professional consulting services to industrial clients. The firm's clients represent a diverse industrial mix from local small firms to large national corporations. They include the glass and automotive, building products, food, and a variety of other industries.

Features of the Site. Provides information on the company, has an online résumé submission capability, and lists career and client information. It provides a number of links of interest to job seekers and recruiting managers.

Interim Services, Inc.:
http://www.interim.com

Description. Established in 1946 and based in Fort Lauderdale, Florida, Interim Services is a major provider of a broad array of staffing classifications with over 1,000 offices worldwide. It employs more than 375,000 full-time and flexible employees. Sales for 1995 were $1.4 billion. It is publicly trades on the NASDAQ. In addition to providing accounting, legal, office, and clerical staff, and physicians and health care workers, Interim provides technical staffing in programming, networking, communications, and databases.

Features of the Site. A very large site with a wealth of information on this firm including corporate description, office locator, a career center, services the firm offers, opportunities for employment, and its own search engine to assist the viewer locate information within the site.

Raymond Karsan Associates:
http://www.raykarsan.com

Description. This is a fast-growing executive search and professional contract employment provider with seven offices across the United States.

Features of the Site. Provides a corporate history and structure, a list of industries served, services provided to clients, career opportunities available by industry, and locator and contact information.

Keane, Inc.: *http://www.keane.com*

Description. Founded in 1965 and headquartered in Boston, Massachusetts, this firm is a software services firm that designs, develops, and manages software for companies and health care facilities. In 1995, Keane derived 85 percent of its revenues from companies it worked with the prior year. With a nationwide network of 40 offices, it provides IS planning, application development, application management outsourcing, help desk outsourcing, year 2000 compliance management and project management, and professional training.

Features of the Site. Provides an overview of the firm and its history, its services, extensive information on employment opportunities, employee benefits and career opportunities, contact information for its 40 offices, and e-mail résumé submission instructions and capabilities.

Kelly Services:
http://www.kellyservices.com

Description. This Michigan-based major provider has more than 1,400 offices throughout North America, Europe, Australia, and New Zealand.

The annual sales exceed $2 billion. A Fortune 500 company providing services of more than 750,000 employees annually.

Features of the Site. A very large site that provides a wealth of information for the job seeker. In addition to corporate information, the site offers an overview of the company and its services. The site provides a location search engine to help the applicant find the nearest office location. The site includes current professional, technical, and scientific assignment opportunities, a résumé maker for electronic submission, and helpful tips on résumé development (with samples) and interviewing. There is an internal search engine so that applicants can look for jobs based on self-selected criteria.

Manpower—Technical:
http://www.manpower.com

Description. Manpower Technical is a division of Manpower, the world's largest staffing service. Currently, it provides staffing needs for close to 90 percent of the Fortune 500 companies and has more than 1,200 offices.

Features of the Site. Provides information and news on Manpower and Manpower Technical. Via an image map, applicants can search for the closest office. The viewer can review available jobs and submit a résumé electronically.

Nesco Service Company:
http://www.nescoservice.com

Description. Nesco Service Company has been a New England–based (Waltham, Massachusetts) human resource provider of technical and information technology professionals since 1960. Nesco serves banking, brokerage, insurance, telecommunications, data processing, computer sciences, utilities, aerospace, and electronics industries.

Features of the Site. Gives basic information on the firm and its areas of expertise and the disciplines it serves. It lists the benefits for its employees and contact information. It does not have an online résumé capability, but there is voice, fax, and e-mail contact information.

Norrell Corporation: *http:// www.norrell.com*

Description. Norrell Corporation, founded in 1965 and based in Atlanta, Georgia, is a leading provider of temporary staffing and out-sourcing services. Through its 400 locations and 6,000 full-time associates, it serves 19,000 customers. In 1995 it placed over 200,000 men and women in short- and long-term assignments. Revenues for the fiscal year ending October 29, 1995, were $813 million. Its common stock is publicly traded on the NYSE.

Features of the Site. Site provides corporate information, franchise, investor and employee information.

Olsten Corporation: *http://www.olsten.com*

Description. Founded in 1950, Olsten Corporation is one of the leading providers of staffing, particularly in home health care. With over 1,300 offices throughout North America, South America, Great Britain, and Continental Europe, the firm serves a large client roster. It provides staffing services at many job levels through Olsten Staffing Services, information technology services through its IMI Systems, Inc. (*http://www.imisys.com*) division, and home health care through Olsten Kimberly QualityCare.

Features of the Site. A large site that provides a wealth of information on this firm and its divisions, job search tips, the skill categories the firm searches for, 1-800-worknow (its U.S. and Canada information line), and a list of offices accepting résumés online. It does provides an online application form through worknow.com feature.

The Pollak and Skan Group: *http://www.pscts.com*

Description. The Pollak and Skan Group is a national contract technical services firm in engineering and technical disciplines. Three companies make up the Pollak and Skan Group: Pollak and Skan, Inc., Contract Technical Services, and P/S Datapro formed in 1987 to provided professionals for information technology and creative associates for contract technical services.

Features of the Site. Provides information on the companies that are part of the Pollak and Skan Group, including the skills and capabilities sought, types of employment available, and benefits offered, and a locator guide for jobs currently available and a company office locator guide.

Quantum Resources®:
http://www.quantum-res.com

Description. Founded in 1963 and headquartered in Richmond, Virginia, this firm provides staff with information technology and other technical skills in over 500 professional classifications to high technology, government, business, and industry. In 1995, through four divisions and 13 offices, it provided over 5,000 employees to over 800 clients in 22 states.

Features of the Site. Provides information on the company and an image map locator for jobs. It has the ability to view jobs online. It does not provide an interactive fill-in résumé, but it does accept e-mail ASCII text submissions.

TAD Resources International:
http://www.tadresources.com

Description. Established in 1956, this Cambridge, Massachusetts–based staffing firm enjoyed 1995 sales of over $1 billion. It has over 350 offices worldwide and provides services to clients in more than 30 countries. TAD provides engineering, design, and drafting professionals in a variety of technical disciplines including electrical, mechanical, manufacturing, and hardware and software. It provides design and development support for the international automotive industry. It supports the computer, telecommunications, and energy industries. TAD has its own accredited postsecondary schools providing technology-specific training.

Features of the Site. Provides corporate information including financial. The site includes a detailed directory of the 350 offices and searchable placement information. Site offers the reader an opportunity to submit a résumé via e-mail.

Talent Tree: *http://ttrec.com*

Description. Founded in Houston, Texas, in 1976, this company now employs more than 95,000 temporary workers annually through its network of 140 offices in 30 states. This firm provides staffing for health care, insurance, financial services, professional services, communications, and retail and distribution. It offers staffing to meet specialized needs such as accounting and legal temporary. Technical staffing is part of the firm's offering, not its primary thrust.

Features of the Site. Provides general information on the company, specific information for employees and potential employees about the company and its benefits, information for clients and potential clients, training programs available with the firm, and industry news. The site provides a listing of current job openings and has an online application capability.

Technical Aid Corporation:
http://www/1tac.com

Description. Founded in 1969 in Boston, Massachusetts, with 120 locations nationwide, this privately held company provides services in 50 state and 21 companies. It is actually a family of companies, each specializing in meeting specific needs. The companies include MicroTemp Systems and Programming, which addresses client software engineering and product development needs; EDP Contract Services, which is a supplier of information technology professionals; TAC Nationwide Staffing, which provides contract professionals with specialized skills; TECH/AID, which provides top-level engineers, designers, drafters, and technicians; and TAC/TEMPS, which offers administrative and office support personnel.

Features of the Site. Gives information on the TAC family of staffing companies, a list of positions currently available nationwide, and an online skills submission form.

TECHSTAFF, INC.:
http://www.techstaff.com

Description. Founded in Milwaukee, Wisconsin, in 1985, this company is a specialty supplier of employment services. With offices in the Midwest, California, and Florida, this firm provides services across a broad geographical area. TECHSTAFF offers contract and permanent placement services for engineers, drafters and designers, information systems personnel, and those with a variety of other technical skills.

Features of the Site. Provides a company history and mission, list of office locations, franchise opportunities, and current job offerings. It has online résumé submission and job-posting capabilities.

TechTemps, Inc.:
http://www.techtemps.com

Description. This firm based in Oak Brook, Illinois, and it provides temporary and a variety of contract service options for information technology, engineering, design and drafting, CAD, and manufacturing and plant management. Through Technical Resources, Inc., it also provides permanent placement services on a contingency, retained, or out-placement basis.

Features of the Site. Provides information on the company and a listing of available opportunities categorized by type of opening. Candidates are urged to submit their résumés via e-mail. Does not provide an office locator guide.

Volt Services Group: http://www.volt.com
see also http://www.volt-tech.com

Description. This firm was founded in New York City in 1950. It has evolved from a primarily southwestern company to a nationwide company with over 120 offices. It is ISO 9002 certified. One of the largest contract staffing providers, this firm provides staff for information systems, engineering and architecture, aerospace, marine, petrochemical, pharmaceutical, and energy industries.

Features of the Site. Gives information about the company and its specialty divisions, a phone guide locator, instructions for how to place an order for an employee, FAQs on temporary employment, a list of benefits provided, and job search information and instructions.

Western Staff Services:
http://www.westaff.com

Description. Founded nearly 50 years ago, Western remains one of the largest staffing firms with over 350 offices in the United States, the United

Kingdom, Norway, Denmark, Australia, and New Zealand. Western provides technical personnel for a broad range of disciplines including information technology, electronics, and engineering. It covers the spectrum of engineering, providing a range of opportunities in architecture, CAD, civil, chemical, electrical, mechanical, and manufacturing.

Features of the Site. Provides industrial, medical, and office as well as technical staff. The site gives the reader information on the company and its divisions and franchise opportunities. It has a locator guide to assist in finding the office. This site lists the types of employment opportunities available with the firm, and has interactive résumé submission capabilities.

Winter, Wyman Contract Services:
http://www.winterwyman.com

Description. Located in Boston, Massachusetts, with an office in Atlanta, Georgia, this firm has a 25-year track record in high-tech recruitment and staffing. The firm's specialty is staffing software engineering and information technology.

Features of the Site. Provides information on the company, its services and clients, and contracting jobs. Job applicants can send résumés and request information on the company. Employees can also use the site to submit hours.

H.L. Yoh Company:
http://www.hlyoh.com

Description. With over 40 offices nationwide, this Rochester, New York, firm is one of the largest providers of contract technical staff. Originally founded in 1940 to provide contract engineering and design personnel, the firm now handles a broad array of technical disciplines skills through its technical, information technology, and scientific divisions. H.L. Yoh now handles computer science, MIS, chemical, mechanical, electrical, electronic, and biotechnical professionals.

Features of the Site. Provides information for the client and the candidate. There is an image map office locator and an online brochure describing the services the firm offers. There is an online résumé builder to help the candidate submit a résumé. It lists career opportunities available with the firm as well as contact information for questions.

Other Employment Resources

The number of resources available on the Internet for job seekers continues to grow. This list is not intended to be all inclusive. It is designed to give the job seeker a start. Many sites give extensive links that will lead to new resources with every mouse click. As with any Internet resource, the reader should be mindful of the dynamic nature of the Internet. Sites are growing and changing. Readers should also be aware that more businesses are adopting profit-generating strategies for using the Internet. Some resources previously indicated as free may impose user fees in the future. The user is urged to exercise usual cautions in using fee-for-service alternatives.

Association for Computing Machinery: *http://www.acm.org*

Description and Features of the Site. This association site provides information for its members and for job seekers with computer technology skills. It provides a digital library, chapter information, and other association directed information. The career section reviews current information available on job seeking and offers links to a variety of other sources.

Career Magazine: *http://www.careermag.com*

Description and Features of the Site. The site is developed and maintained by National Career Search (NCS), an agency that offers services to employers and recruiters. The site daily downloads and indexes job postings from the major Internet newsgroups. The postings are searchable by location, job title and/or skills, needed. For the human resource professional it provides an easy to use tool for searching résumés banked with the *Career Magazine.* Job seekers can bank their résumé with the site for a fee via an online form. They can also review information on employers. At the same site, the job seekers will find products, services, and news (including job fairs) for career management as well as numerous links to other career-enhancing sites. The Career Forum section of the site provides a moderated discussion area for networking.

CareerMosaic:
http://www.careermosaic.com

Description and Features of the Site.   This heavily graphic, hence slow-loading, site run by Bernard Hodes Advertising, Inc., provides a wealth of information for the job seeker. It includes the j.o.b.s—jobs offered by search—database to link job seekers to job openings by job description, title, company, and address. It has information on online job fairs, an international gateway, and a career resource center for finding helpful information on job hunting. It offers for college students its College Connection, and for health care professionals its Health Care Connection. For human resource professionals, there is an HR Plaza that is specially designed to meet their needs.

Career Path.com:
http://www.careerpath.com

Description and Features of the Site.   This site allows the reader to search the employment ads from newspapers in 19 major cities. Users must register to use the site beyond the sampling. The service is free and provides daily access to the want ads of major newspapers including *The Boston Globe, The Chicago Tribune, The Los Angeles Times, The New York Times, The San Jose Mercury,* and *The Washington Post.* When checked in early autumn 1996, the weekly listing included 110,948 jobs.

CareerSite: *http://www.careersite.com*

Description and Features of the Site.   This site is the creation of Virtual Resources Corporation of Ann Arbor, Michigan. It provides candidates and employers the technology to interactively search for either job opportunities or for job seekers with desired credentials. Job seekers can submit their résumé for free and then wait for the site's virtual agent to notify them of opportunities that match their criteria. No material is released to a potential employer without the prospect's permission. This ensures up-to-date information. Employers who pay to list their openings with this service gain the benefits of the site's virtual recruiter. This will match individuals with openings.

Contract Employment
Weekly Online: *http://www.ceweekly.com*

Description and Features of the Site. This site is a service of C.E. Publications, publishers of *Contract Employment Weekly*, a publication that lists job openings for contract technical employment. The site maintains separate searchable databases for subscribers and nonsubscribers. The paper version is a weekly newsletter sent first-class to subscribers. It provides information about immediate and anticipated technical job openings throughout the United States, Canada, and overseas. This site provides a number of services of interest.

E-Span: *http://www.espan.com*

Description and Features of the Site. This site is a virtual online classified. It provides the job seeker the opportunity to search a position using keywords or other criteria. For the human resources professional looking to fill a position, the site provides an economical access to reach thousands of potential candidates. There is also a section of the site dedicated to providing support for human resource professionals.

ITTA: *http://www.it-ta.com*

Description and Features of the Site. This site is maintained by an association of IT professionals. Included in the site are member services and enrollment information as well as job hunting information. The services, which include résumé posting and automated job reply services, are available to members only. There is also news and information of interest to IT professionals. The focus is on consulting and contracting.

Job Web: *http://www.jobweb.org*

Description and Features of the Site. This site is sponsored by the National Association of Colleges and Employers. It provides a number of services of interest to job seekers and employers including job listings, job search and industry information, resources for career planning and materials of interest to human resource professionals. Employers can also post jobs with the site. It is a resource-rich site.

The Monster Board:
http://www.monster.com

Description and Features of the Site. This site is run by Adion Information Services, a New England agency. It is an extensive site for job seekers offering job listings, résumé submission services, and information on employers and career events of interest to technical (and other) job seekers. Although comprehensive in its coverage, it is particularly directed to New England job seekers. It also provides specialized services for those either just coming out of college or beginning their careers.

National Association of Computer Consultant Businesses:
http://resourcecenter.com/rc/owa /jba.search

Description and Features of the Site. This association-maintained site provides a NACCB Job Board that posts technical jobs and contract employment opportunities currently available. The association also maintains a résumé bank and a mailing list. The site has résumé submission capabilities.

National Association of Temporary Staffing Services:
http://www.natss.org

Description. NATSS represents more than 1,400 staffing companies that operate approximately 9,000 offices throughout the United States. From it original goal of ensuring that competent temporary help services are available to business while providing flexible employment opportunities for the workforce, NATSS has evolved to serve the entire staffing industry.

Features of the Site. This site is a source of information and publications on the staffing industry. It provides a number of databases of interest to those involved in the staffing industry. It posts a calendar of events of interest to those in the staffing industry and information on the organization, the state chapters, and other membership information.

National Technical Employment Services Association: *http://www.ntes.com*

Description. This site is maintained by NTSA, a nonprofit trade association comprising companies that provide a broad range of services to industry and government. NTSA member firms provide contract staffing to a wide variety of businesses and government entities. NTSA members employ more than 280,000 technical services personnel and generate over $5 billion in sales.

Features of the Site. The site provides a central résumé database to which applicants can download their own résumé, job listings, a recruiter connection to link job seekers with contracting firms, links to recruiter home pages, and other job hunting services. It provides information on technical temporary staffing and limited information on the labor market.

Net-Temps: *http://www.net-temps.com*

Description and Features of the Site. This site is a commercial venture linking those recruiting temporary personnel with available applicants. The recruiting agencies pay a fee to subscribe to the service. This permits them to post their offerings, maintain a company home page, and e-mail services. The services of this site are free to the job seeker. An open text search engine enables viewers to match their work experience, education, location preferences, and career interests with available listings. The agencies listed in this site do not typically represent those large agencies with enough name recognition and mass of openings to maintain their own sites.

Software Jobs Home Page: *http://www.softwarejobs.com*

Description and Features of the Site. This site is administered by Allen Davis & Associates, a search firm that specializes in software and information technology (IT) search and placement. The site has grown out of the firm's BBS. The firm maintains a network of over 600 recruiting affiliates in almost every state. The firm has four key areas of specialization and publishes an online newsletter on the site: Windows/GUI Opportunities,

RDBMS Careers, Opportunities in SAP, and the Independent Consultant. The site has a special area for contracts and gives the viewer an opportunity to search the software jobs database, view new additions, or apply for the positions. There is ample information to assist the hiring manager as well as a number of career-related resources and useful links of interest to software job seekers.

Appendix: Building the Job Order

This appendix is designed to provide examples of the methodology recommended for constructing a comprehensive job order. Exhibit A-1 illustrates the process.

To Eliminate Job Order Confusion, Start with the Job, Not the Existing Job Description

Determine Required
Duties & Tasks

List Relevant
Knowledge, Skills
and Experience

Select Relevant
Parts of Existing
Job Descriptions

Build the
Job Order

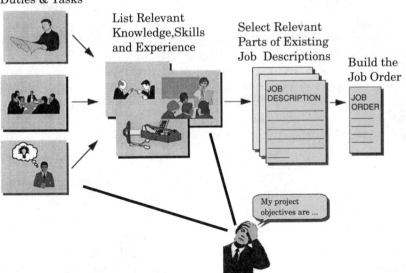

Exhibit A-1. To eliminate job order confusion, start with the job, not an existing job description.

Step 1. Establishing the Hiring Manager's Objectives

Get a complete description of the hiring manager's objectives for filling a contingent position and the work that must be done to complete the assignment that generated a need for contingent staffing. If the work is a project, the description should be strictly limited to the tasks that must be done by the contingent hire to bring the project to successful conclusion. If the work is more open-ended in nature, the description should reflect this. In both cases, the person seeking the description should guard against unnecessary expansion of required tasks, or "scope creep."

Descriptive information is usually generated in a series of interviews. Interview candidates include the hiring manager or managers, potential coworkers (some of whom may be responsible for the same or similar work), and individuals who may be using the work generated by the contingent hire.

Single-source interviews, while less time-consuming, do not always provide a complete picture of the work or level of performance that will be expected when the position is filled. Resolving conflicts in expectations at this point in the staffing process is dramatically less painful than dealing with them after the new hire is on board. In addition, interviews with individuals who have different views of the work can be instructive. For example, coworkers who are doing similar or same work are the best source of information on how frequently different tasks or activities must be done.

Supervisors or managers, on the other hand, have a unique perspective on why individuals succeed or fail at certain tasks. For that reason they are usually well represented at one form of group interview, the *critical incident meeting*. At this meeting, participants provide concrete examples of effective and ineffective performance. From these descriptions, the interviewer can build a clear picture of the overall standards of success for the work at hand and can represent these standards in discussions with temporary staffing agencies.

Step 2. Establishing Minimum Levels of Expertise and Experience

Determine the minimum knowledge, skill, and/or experience level required to successfully complete the tasks and activities that were agreed to in step 1. The word "minimum" is used deliberately to combat a common problem in perception that grows in the minds of incumbents and managers. Experienced human resource professionals would agree that long-term incumbents, regardless of level, are convinced that their jobs require more skills and experience than any other job in the company. If not questioned, these assertions lead to job descriptions that specify "10 years of experience and a master's degree in engineering" for entry-level jobs! Therefore, it is important to discover the minimum skill and experience requirement for each component of a job. For example, a job that requires knowledge of algebra and some knowledge of calculus requires just that, algebra and calculus. The job does not necessarily require a B.S. in engineering, even though algebra and calculus are required courses in most engineering programs. The same issue arises in the area of experience. We often find that *years* are specified, when what is important is the *number of relevant experiences* that an individual may have had. In stating skill requirements, the interviewer needs to be aware of one significant difference between permanent and temporary employees, namely, what they will be expected to learn prior to being an effective contributor. In the case of permanent employees, the employer may be willing to make a significant investment in training, thereby increasing the value of human assets in the organization. This is based on the expectation of a long-term payout. This is absolutely not what is expected in the case of contingent hires. They are expected to be effective on day 1 of employment, with only a few exceptions.

One of those exceptions that deserves special treatment in the skill identification exercise is how to address skills that are unique to the hiring organization. For example, the hiring company may have proprietary software or manufacturing processes that are used to perform certain tasks. It would be unrealistic to expect an outside hire to have experience with these tools or processes.

In these situations the interviewer needs to determine what foundation skills were present in the incumbent population that allowed them to become effective users of the proprietary processes or systems. In the critical incident meetings, supervisors and managers need to indicate what differentiated individuals who were more or less successful in achieving rapid proficiency in these areas.

A second exception to the "expectation of a fully trained employee on day 1" rule is present when an employer is entertaining the possibility of a "contract-to-permanent" relationship. In these cases, the employer may wish to add skills to the job order that go beyond the immediate task requirements. In addition, the employer may be more willing to invest in training for the contract employee.

Step 3. Grouping the Job Tasks

After minimum required skills, knowledge, and experience are linked to required tasks, the tasks can be logically grouped. Sometimes the tasks can be grouped on the basis of similar skill, knowledge, or experience requirements. In other cases, tasks fall into natural groups, which may already be present on existing job descriptions or summaries.

It is at this point in the process that the person preparing the job order can logically turn to the preexisting job descriptions and summaries for guidance and language. Since the scope of the work to be done has been delimited by the tasks and the skills requirements are linked to the tasks (unless the contract-to-permanent exception applies), the job description or summary will serve more as a filter than as a guide for the creation of the job order. An example of this process is provided later in this appendix.

Step 4. Prioritizing the Job Tasks

Prioritize the tasks and related skills in order of criticality to guide the staffing agency in candidate recruitment and selection.

Step 5. Constructing the Job Order

Using the task groups and related skill requirements as a foundation, construct the job order. In addition to the foundation, the job order should include other important factors that will likely narrow the field of candidates to those that will prove successful in the specific work environment. Exhibit A-2 shows some of the factors that should be included in the job order.

Among the factors that influence the relative success of contract hires, none is more important than cultural fit. A clear statement describing the work environment, expectations of coworkers, quality standards, physical environment, and rules of conduct should round out a job order. Examples include smoking rules, team participation, and work space. The hiring manager should review the discussion on cultural fit elsewhere in this volume and select those items that are relevant for inclusion. In addition, the job order should include other essential screens, such as required security clearances, drug testing, and physical requirements (consistent with the Americans with Disabilities Act compliance).

Although Tasks, and Related Knowledge, Skills and Experience Provide the Foundation of the Job Order, The Finished Document Reflects Other Significant Factors

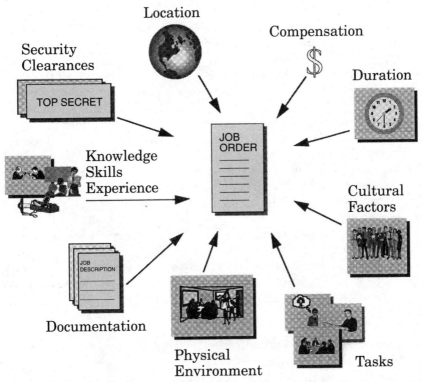

Exhibit A-2. Although tasks and related knowledge, skills, and experience provide the foundation of the job order, the finished document reflects other significant factors.

The Methodology at Work: An Example

In the wake of a series of articles naming their region as one of the top three U.S. retirement locations, a five-county area in a southwestern state is in the throes of a building boom. The gas and electric public utility serving the area has experienced a rapid increase in demand for residential and small commercial services. As a result, the head of engineering services for the public utility has several projects in his backlog that cannot be addressed by the existing technical workforce during the next 12 to 16 months. He needs help, and he needs it now!

After reviewing one of the bigger residential jobs, he makes a brief list of tasks, pulls out several yellowed job descriptions from his files, and begins to match the tasks to the existing jobs. Given what he knows, he calls the recruiting manager in human resources and asks her to begin a search for an engineer (see job description 1). He is aware that some of the tasks that need to be done on the project will require little more than drafting, but he feels that an engineer will be sure to have all the skills he needs and more. Better safe than sorry.

The recruiting manager informs the head of engineering services that due to a hiring freeze, she cannot recruit a full-time engineer. However, she offers to contact a contingent staffing firm and get the engineering director some temporary relief, at least until the residential project is through. She schedules an appointment with the head of engineering services, and asks him to provide her with the names of other people in the department who either *do* similar work or who directly supervise those who *do*.

The next morning the recruiting manager meets with the head of engineering services. First, she has the hiring manager outline his objectives for the position and then has him clearly outline what a successful outcome on the project would look like. Then she asks him for a list of tasks that need to be done to complete the project. The list includes the following:

a. Meet with construction contractors and get site plans for the new residential development

b. Do field investigation and inspection to determine what structural interferences must be addressed in providing new gas and electric services

c. Prepare sketches for construction drawings

d. Prepare construction drawings based on the input of the engineering manager-in-charge for the project

e. Review drawings for regulatory compliance, including safety and environmental standards

f. Handle interpersonal issues with contractors and resolve problems

g. Solve engineering problems

h. Control project documentation

i. Complete project work in 3 months from date of hire

Next, the recruiting manager meets with an engineering manager, a recently hired engineer, an engineer with 3 years of experience, and a design technician (see job description 2). By lunchtime, she has reviewed the list with all four and is ready to discuss the job with the head of engineering services.

First, she reviews the task list with him, confirms the assignment duration, and provides him with the information gleaned from his subordinates. It appears, she says, that some of the tasks clearly match parts of the engineer job description. In addition, some of the tasks are found in several other job descriptions, most notably that of the design technician. Because they will not be hiring someone to fill a permanent job, they can build a hybrid description that allows them to seek someone who can do exactly what they need. It will not be necessary to hire someone who meets all the qualifications of either the engineer or design technician classifications.

The head of engineering services is skeptical. How can I be sure that I will get exactly what I need if I don't get a graduate engineer? At this point, the recruiting manager brings out the task list and her notes from the interviews and volunteers to provide the hiring manager with a point-by-point review of the tasks versus specific language in the job description.

Speaking directly from the list prepared by the head of engineering services, she points to tasks a and f and asserts that these can be grouped into the category of "technical interface with contractors and other customers." A possible skill match for these tasks can be found in design technician qualifications II.6 and II.7. To make certain that she is on target, she asks if there are any types of relationship skills needed beyond those specified. After gaining agreement on the first point, she proceeds to tasks b, c, and d, which are partial or complete matches with design technician duties I.1, I.2, and I.4. The head of engineering services asserts that task d, preparation of construction drawings…, exceeds the qualifications of the design technician and may actually be more accurately reflected in the work of a design engineer (job description not shown). The recruiting manager agrees and changes her notes accordingly. At the same time, the recruiting manager probes more deeply into task d. She asks if the work will be done at a drafting table or using one of several CAD systems that have been purchased by the department. When the hiring manager indicates that he would expect the contract employee to be CAD proficient, the recruiting manager has him specify the system and software version to be used and the types of drawings that the individual would be expected to produce (primarily 2D, with some occasional 3D wire frames that can be drawn from a preexisting, online library of objects).

The recruiting manager notes that tasks e and h are direct matches for parts of the engineer job description, namely, duties A.4 and A.6, respectively. After getting agreement, she addresses item g, which is very general. From her notes, she indicates that the subordinates that she interviewed were also curious about the engineering problems that were to be solved by this individual. With characteristic candor, the head of engineering explains that he put this qualification on the list to let human resources know that he wanted a graduate engineer. In the discussion that follows, the recruiting manager and the head of engineering services talk frankly about both the economics of the situation and the concerns that the hiring manager has concerning contract hires. In the end, the hiring manager agrees to a skill and experience mix that more closely approximates the design technician, with the qualification that the individual have actual experience in the utility industry.

After making certain that the list of requirements is complete, the recruiting manager asks the hiring manager to specify what percentage of time would be devoted to specific activities and then to prioritize the tasks and skills on a 3-point scale. The priorities will be communicated to the contract staffing firm and will help guide the candidate selection process. At this point, the head of engineering services asks if the recruiting manager has enough information to proceed. The recruiting manager offers the hiring manager two options. One is that she leave a checklist with the hiring manager on which he can provide a description of the working environment, the geographic area in which the work is to be done, any special certifications required (licenses, CDL, etc.), work rules, quality standards, and the cultural norms of the organization. As an alternative, she is willing to fill in the checklist with him. In half an hour, the work is done, and the recruiting manager is ready to complete the job order with the assurance that the staffing firm will have clear directions for staffing the job.

As a follow-up exercise, the recruiting manager sets up an appointment with the hiring manager, reviews his experience with the process, and offers to prepare him to complete the next job order on his own.

Job Description 1: A Job Summary for an Engineer in the General Engineering Department

Classification: Engineer

I. *Function*

Under general supervision, an engineer is responsible for performing the following duties commensurate with experience and qualifications.

A. Duties

1. *Project Management*

All activities associated with management of multidiscipline projects as well as projects in a particular field of engineering. Responsibilities include scheduling, specifications, procedures, procurement, inspection, installation testing, cost control, and regulatory compliance.

2. *Research*

Obtaining, reviewing, and organizing field data and theoretical engineering and economic information relevant to a particular problem or opportunity including preparing detailed reports and recommendations for management review.

3. *Design*

(a) *Sponsor engineer:* Performing and directing the calculations necessary to provide performance data for the layout and detailing of specialized equipment and material installations, including computer modeling and system simulation.

(b) *System protection engineer:* Designing protective relaying and control equipment for transmission lines, distribution circuits, and substation and power plant equipment emphasizing safety for company personnel and the public, providing a high degree of reliability of service, and minimizing damage to system components during equipment failures.

4. *Testing*

Designing and directing test procedures to establish equipment performance levels. Writing detailed reports on procedures and data acquired.

5. *Construction supervision*

Directing and sanctioning construction and installation of equipment and systems, initiating field modifications, and controlling schedules, costs, and regulatory compliance.

6. *Regulatory Compliance*
Obtaining and reviewing all applicable regulatory information regarding materials, performance, safety standards, and environmental restrictions affecting a particular equipment installation or construction project.

7 *Basic Engineering*
Directing and performing computations, studies, drafting, economic analysis, system performance evaluations, specification preparation, cost justifications, preparing procedures and standards, drawing and invoice review and approval, scheduling and administering meetings, preparing detailed reports and recommendations, and utilizing problem and decision analysis techniques.

8. *Directing Engineering Personnel*
Directing the work of other engineering personnel including engineers' draftsmen, cooperative engineering students, and others as assigned.

9. *Documentation*
Organizing, compiling, and documenting all engineering work associated with assigned projects assuring complete and accurate records, correspondence, and drawings.

10. *Expediting and Coordination*
Expediting and coordinating equipment being procured, construction and installation schedules, equipment outage schedules, and system availability requirements while controlling project costs, quality, safety, and reliability.

11. Management Presentation
Preparing appropriate reports and documents for management review as assigned.

12. *Miscellaneous*
Performing miscellaneous engineering functions including assisting customers, training and consulting with other engineers inside and outside of the company, and assisting in handling programs to attain company objectives. Representing the company and profession as required.

II. *Qualifications*
Must meet the company's *general qualifications* and, in addition, must:
   A. Have a bachelor of science degree in engineering from accredited engineering college
   B. Have at least 1 year's experience as a practicing associate engineer

   C. Possess an engineer-in-training certificate

   D. Have demonstrated the ability to perform the duties of an associate engineer

   E. Be capable of interacting with coworkers, contractors, and/or manufacturer's representatives in a professional manner utilizing both oral and written skills

   F. Exercise judgment in all phases of the work

   G. Demonstrate the ability to plan, design, schedule, and direct project work, in the field and in the office

   H. Possess general knowledge of various engineering phases applicable to the utility industry

   I. Be capable of dealing with others in professional manner within the scope of assigned responsibilities

   J. Perform neat and accurate work

   K. Possess general knowledge of the capabilities of various data-handling and computing systems as they may apply to this classification

   L. Demonstrate creativity in problem solving

   M. Communicate effectively, in writing and orally, with all levels of personnel at various technical levels

   N. Pursue technical and/or managerial development after graduation and maintain competence and knowledge of advancements in their chosen engineering discipline

Job Description 2: A Job Summary for a Design Technician in the Engineering and Planning Department

Classification: Design Technician

I. *Duties*

Under general supervision, performs normal and routine design work requiring individual action, judgment, and decisions. Provides direct field assistance or inspection during construction of gas and electric facilities as needed and performs such duties as:

   1. Making field investigations and inspections for the determination of location of facilities, interference with other structures, negotiations for final arrangements for facility construction for new business, system betterment, and construction ahead of public improvements

   2. Preparing sketches for contracts, construction drawings, and permits

3. Preparing estimates for work order preparation and bills of materials for the coordination of purchasing, stores, and construction, including the dates of construction
4. Coordinating drawings, sketches, and bills of materials with gas and electric operating department for suggestions and approval
5. Supervising the drafting work of others
6. Preparing and maintaining construction unit costs
7. Performing similar or less skilled work

II. *Qualifications*

Must meet the company's requirements as to general qualifications and, in addition, must have all the qualifications of a drafting technician A and, in addition, must:

1. Have an associate's degree in engineering technology from an accepted college
2. Have had at least 2 years' experience as a drafting technician A or the equivalent
3. Have the ability to plan and schedule project work in the field
4. Be able to make field studies and surveys
5. Have a general knowledge of the various engineering and planning phases of the public utility industry
6. Be capable of dealing with operating department supervisory personnel, other utility engineering and operating personnel, public authorities, and construction contractors
7. Be able to exercise judgment relative to the decisions necessary in the work
8. Be capable of neat and accurate work

Index

AT&T, 2
AccuStaff, 212
Adecco, 210
Aerotek, 210
alliances, (strategic) for staffing, 163–167
Americans with Disabilities Act (ADA), 112, 232
anemia, managerial and technical, 106
appraisal, performance, 154, 179, 194
Association for Computing Machinery, 223
Autocad, 29, 180

benching, 76
boundary spanning roles, 105
Butler International, 211

C++, 78
CDI Corporation, 211
career ladders, 179
Career Magazine, 223
career management, 80, 82
  self-directed, 89, 90
  setting goals for, 91
CareerMosaic, 224
Career Path, 224
CareerSite, 224
career worksheets, 27–32
chains, supply and distribution, 159
changes of control, 170
citizenship, verification of, 153
Cobol, 88
communications procedures, 130
compensation, 74–76, 90, 179
  skill-based, 13
computer-aided engineering (CAE), 147
Computemp, 213
Computer Horizons Corp., 213

Computer Task Group (CTG), 212
contingent employees:
  as percentage of workforce, 4
continuing education, role in career planning, 88
Contract Employment Weekly Online, 225
contract staff:
  advantages of using, 100
  coaching of, 203
  cost-benefit of, 105
  dehumanizing of, 124
  drawbacks of using, 103
  false promises and, 126
  impact of, 130
  job descriptions for, 128, 152
  loyalty of, 150
  orientation procedures for, 129, 131
  over managing, 132
  performance standards for, 125
  personnel policies for, 128
  remediation needs of, 203
  social functions, at, 133
  teams and, 135–139
  training, 132
  trust and, 133
  use in sales forces, 105
  work rules and, 202
contract staffing firms:
  account executives of, 148
  advantages of working for, 97–98
  applying for employment with, 65, 66
  as validators of experience, 16
  billing rates of, negotiating, 167
  confidentiality agreements and, 73
  employee orientation with, 74
  evaluating, for potential employment, 64, 68–70
  finding on the Internet, 64
  hiring process of, 24, 25

contract staffing firms (*Cont.*):
  interviewing with, 65–67, 71
  job offers with, evaluating, 74
  job orders and, 181–190
  labor disputes and, 169
  negotiating contracts with, 167
  orientation for employees of, 152
  sexual harassment and, 172
  subcontracting between, 168
  working for,
    compensation and benefits, 74–76, 90
    eligibility requirements, 75
    during labor actions, 73
    holidays, 73
    length of contract assignments, 64
    retirement plans, 401 (k), 74, 76, 90
    sick time with 73
consultants, 95
contracts with staffing firms:
  negotiating, 167
contractors, independent, 95, 97
  20-point test, 113–114
  IRS and, 96, 113–114
  legal issues, 112
critical incident meeting, 230, 231
CTS Technical Services, Inc., 212
culture, corporate, 186, 232
  of entitlement, 187

DLD Technical Services, Inc., 214
Davidow, William H., 5
Department of Labor, 81, 113
diversity, of workforce, 119
  tolerance of, 186
drug testing, 153
Dunhill Personnel System, Inc., 214

E-Span, 225
education, verification of, 153
employee handbooks, 73, 153
Employee Retirement Income Standards
  Act (ERISA), 112, 134
employees:
  as assets, 85
  as human capital, 86
  empowerment of, 119
  leasing, 109
    advantages of, 111
    disadvantages of, 111
  temporary, 131

employees (*Cont.*):
  training of, 169
employer of record, 134, 171, 202
employment, verification of, 153
employment models
  free agency, 11, 16
  shamrock, 4–5, 11
employment tactics, nontraditional,
  93–94
Engineering Corporation of America,
  214
engineers:
  job preferences of, 12
entitlement, culture of, 187
environmental engineering, growth of
  industry, 87
evaluation, performance, 194, 196

FICA, 96
Federal Labor Standards Act (FLSA), 112,
  114, 179
financial planning, need for, 90
flexible workforce, 102
Ford Motor, 155
Fortran, 88

General Motors, 2
Generation X, 120, 121
glass ceiling, 3

Handy, Charles, 4
hiring, costs of, 101

IBM, 2, 80, 118, 155
IMI Systems, Inc., 215
ITTA, 225
interdependencies, of businesses:
  financial, 161
  strategic value of, 159
  value-added, 160
Interim Services, Inc., 215
interviews:
  critiquing, 58
  follow up after, 58
  handling weaknesses in, 55
  panel, 58
  preparing for, 41, 53–57
  telephone, 65–66, 67

Internal Revenue Service (IRS), indepen-
    dent contractors and, 96, 113–114
    misclassified employees and, 134
ISO 900X, 14, 134, 140, 155
ITS Technologies, 24, 148, 173, 201, 203,
    215

job:
    analysis, 181
    description, 145, 179–181
    family, 179
    order, 145, 182–185, 188, 196, 229
    summary, 179–181
Job Web, 226
joint employers, legal obligations of, 110
just-in-time (JIT), 160

Karsan Associates, Raymond, 216
Keane, Inc., 216
Kelly Services, 216

labor disputes, 169
labor market, 145, 200
layoffs, indicators of potential, 83
learning organizations, 89
leasing, employees, 109
    advantages of, 111
    disadvantages of, 111
letter(s), cover, 50–52
    e-mail, 52
    length, 52
    letterhead, use of, 52
    signing, 52
letter(s), thank you, 52–53

Malone, Michael S., 5
Manpower-Technical, 217
manpower planning, 179
materials handling, growth of industry,
    87
Merck, 155
Monster Board, The, 226

National Association of Computer
    Consultant Businesses, 226
National Association of Temporary and
    Staffing Services (NATSS), 63, 226

National Labor Relations Act, 112
National Technical Services Association
    (NTSA), 62, 63, 155, 227
Nesco Service Company, 217
Net-Temps, 227
Norrell Corporation, 218
nuclear family, 82

objectives, defining, 197
Olsten Corporation, 218
organizations:
    total quality, 139
    virtual, 11
orientation, employee, 74, 129, 131,
    152–153, 166, 169
orientation programs, cost of, 104
outsourcing, 106–108

payrolling, 110, 170
performance appraisal, 154, 179, 194
performance evaluation, 194, 196
Pollak and Skan Group, The, 218
professional networks, using, 146

Quantum Resources, 219

references, verification of, 153
resource guide, 21
resume(s):
    accuracy in, 37
    as career story, 34
    career summaries in, 40
    copies of, 42
    data gathering for, 27
    development process, 26
        emotional issues, 34
        psychological barriers to, 26
    electronic retrieval of, 20–22
    electronic submission of, 20
    format:
        acceptance of, 45
        chronological, 43, 44–45
        engineering skill, 49–50
        functional, 43, 46, 47–48
        acceptance of, 46
        uses of, 46
    gaps in, 36
    hot words for 37–40

Internet submittal of, 21
  length, appropriate, 36
  mailing of, 43
  personal data in, 42
  proofreading, 42
  puffery in, 32
  references, listing in, 41
  salary information in, 41
  truth-testing, 43
  use of action words in, 36
retirement plans 401 (k) 74, 76, 90
  need for, 90
review, annual, 195
roles, boundary spanning, 105

scope creep, 230
sexual harrassment, 172
smoking, in workplace, 72, 186
Software Jobs Home Page, 228
span of control, 119
supervisor, on-site, 163, 172
suppliers, just-in-time, 160

TAD Resources International, 219
Talent Tree, 220
Technical Aid Corporation, 220
TECHSTAFF, Inc., 221
TechTemps, Inc., 221
teams:
  absorptive, 139
  contract staff and, 134, 135–139
  cross-functional, 136
  departmental, 134
  development stages of, 136–139
  natural teams, 135
  process, 135
  self-sealing, 139
telecommunications, 145
  growth of, 87

temp-to-perm employment, 94, 103, 154
testing:
  drug, 57, 153
  psychological, 57
total quality management (TQM), 139

underemployment, 94
union contracts, 179
Unix, 88, 145

vendor-on-premises, 171
Virtual Corporation, The, 5
Volt Services Group, 221

wellness programs, employee, 8
Western Staff Services, 222
Winter, Wyman Contract Services, 222
working environment, evaluating prior to employment, 71–73
workforce, composition of, 185, 186
work groups, self-directed, 140
workplace:
  rules in, 73, 169, 196, 202
  smoking in, 72
workforce:
  diversity of, 119
  single-parent families in, 81
  stability of contract labor, 151
  two-career families in, 81
  women in, 81

Yoh Company, H.L., 222

Xerox, 155

About the Authors

AMANDA G. WATLINGTON is Assistant Professor of Business and Marketing at Terra Community College in Fremont, Ohio. She is a consultant in business strategy development, Total Quality Management, and marketing with the Internet. Ms. Watlington has written numerous articles on evolving business strategies and business solutions, and lectures on TQM and the applications of new technologies in business and education.

ROGER L. RADELOFF is President and CEO of ITS Technologies, Inc., which provides technical and professional staffing, consulting, and outsourcing services to major corporations. A registered professional engineer in Ohio and California, he has been involved with many engineering firms as owner, principal, and consultant. Mr. Radeloff received the Small Business Administration's Ohio Business Innovator of the Year Award in 1987, was a nominee in 1994, and was a finalist for the 1996 Entrepreneur of the Year Award.